Frontiers in Clinical Drug Research- Diabetes & Obesity

(Volume 7)

Edited by

Shazia Anjum
Institute of Chemistry
The Islamia University of Bahawalpur
Bahawalpur
Pakistan

Frontiers in Clinical Drug Research – Diabetes and Obesity

Volume # 7

Editor: Shazia Anjum

ISSN (Online): 2352-3220

ISSN (Print): 2467-9607

ISBN (Online): 978-981-5123-58-6

ISBN (Print): 978-981-5123-59-3

ISBN (Paperback): 978-981-5123-60-9

First published in 2023.

Bentham Science Publishers Pte. Ltd.
80 Robinson Road #02-00
Singapore 068898
Singapore
Email: subscriptions@benthamscience.net

BENTHAM SCIENCE

CONTENTS

PREFACE

Causes of diabetes are so complex but type II diabetes is directly linked with obesity at any time of your age particularly if you have excessive fats around your tummy. Obesity triggers your body's metabolism which causes fat tissues to release fats into your blood that directly affect insulin response and make you insulin sensitive- that's why obesity causes prediabetes. In order to understand this complex pathological disorder in our body and thereafter solutions to surpass this challenge, volume 7 of our eBook series is dedicated to cutting-edge articles on Diabetes or Obesity. The update can be found in the first chapter of this present volume that non-enzymatic glycation of proteins, lipids, and fatty acids speeds up due to persistent hyperglycemia and eventually causes associated secondary complications in diabetes.

The authors in the second chapter describe the current strategies of new drugs for diabetes management. For example, the development of novel therapeutic groups such as amylin analogs, incretin mimetics, GIP analogs, active peroxisome proliferator receptors, and dipeptidyl peptidase-4 inhibitors and as well as bioactive compounds from herbs.

The third chapter of this series deals with the debatable topic of using aspartame for T2D patients. More research is still needed to establish the pathological role of aspartame use in T2D.

Chapter four of this volume covers the research to investigate the psychological characteristics and adherence of children and adolescents with Type 2 diabetes. A joint venture of the Faculty of Medicine and Faculty of Arts has developed that mapping mental health and various therapeutic procedures, as well as their positive and negative effects, are of paramount importance for both diabetes and obesity.

The fifth chapter of this volume is about the clinical trials of new-generation anti-diabetic and lipid-lowering agents that also have simultaneously cardioprotective effects.

The sixth chapter of this volume describes that the kidneys are a vulnerable target of diabesity. In this chapter, the epidemiology, pathophysiology, and treatment of diabesity–induced kidney disease are discussed. The special focus on the therapeutic targets and pharmacological management of diabesity-related kidney diseases is described herein.

I hope that the current volume of this series will provide updated information about the recent developments in Diabetes & Obesity treatment for interested researchers and pharmaceutical scientists. I would like to thank the editorial staff, particularly Mr. Mahmood Alam (Director Publications) and Ms. Asma Ahmed (Senior Manager Publications) for their dedicated efforts and the hard work.

Shazia Anjum
Institute of Chemistry
The Islamia University of Bahawalpur
Bahawalpur
Pakistan

List of Contributors

Ahmed Sayed	Department of Internal Medicine, Faculty of Medicine, Ain Shams University, Cairo, Egypt
Anas Al-Refaei	Department of Internal Medicine, Faculty of Medicine, Ain Shams University, Cairo, Egypt
Ahmed Elkeraie	Kidney and Urology Center, Alexandria, Egypt Faculty of Medicine, Alexandria University, Alexandria, Egypt
Arbind Kumar Choudhary	Department of Physiology, All India Institute of Medical Science (AIIMS) Raebareli, Uttar Pradesh (U.P.), India
Beáta Erika Nagy	University of Debrecen, Faculty of Medicine, Institute of Pediatrics, Pediatric Psychology and Psychosomatic Unit, Hungary
Brigitta Munkácsi	University of Debrecen, Faculty of Medicine, Institute of Pediatrics, Pediatric Psychology and Psychosomatic Unit, Hungary
Farah Khan	Department of Biochemistry, School of Chemical and Life Sciences, Jamia Hamdard, Hamdard Nagar, New Delhi 110062, India
Hayat Ullah	Department of Chemistry, University of Okara, Okara 56300, Punjab, Pakistan
Hany Sawaf	Cleveland Clinic Foundation, Cleveland, OH, United States
Issa Haddad	Michigan State University, East Lansing, MI, United States
Jamshed Haneef	Department of Pharmaceutical Chemistry, School of Pharmaceutical Education and Research, Jamia Hamdard, Hamdard Nagar, New Delhi 110062, India
Jasmin Abdeldayem	Department of OB/GYN, Texas Tech University Health Sciences Center, El Paso, TX, USA
Km Neelofar	Department of Biochemistry, School of Chemical and Life Sciences, Jamia Hamdard, Hamdard Nagar, New Delhi 110062, India
Karolina Eszter Kovács	Faculty of Arts, Institute of Psychology, Department of Pedagogical Psychology, Hungary
Khaled Moustafa	Department of Internal Medicine, Faculty of Medicine, Alexandria University, Alexandria, Egypt Faculty of Medicine, Alexandria University, Alexandria, Egypt
Maliha Sarfraz	Department of Zoology, Wildlife and Fisheries University of Agriculture Faisalabad Sub Campus, Toba Tek Singh 36050, Pakistan
Mamoona Noreen	Department of Zoology, Wildlife and Fisheries University of Agriculture Faisalabad Sub Campus, Toba Tek Singh 36050, Pakistan
Misbah Ullah Khan	Center for Nano-Sciences, University of Okara, Okara 56300, Punjab, Pakistan
Mohamed E. Elrggal	Kidney and Urology Center, Alexandria, Egypt
Mohamed Hassanein	University of Mississippi Medical Center, Jackson, MS, United States
Nicholas Elias	Department of Internal Medicine, Morristown Medical Center, Atlantic Health System, Morristown, New Jersey, USA

Omar M. Abdelfattah	Department of Internal Medicine, Morristown Medical Center, Atlantic Health System, Morristown, New Jersey, USA Department of Cardiovascular Medicine, Cleveland Clinic Foundation, Cleveland, OH, USA
Rahman M. Hafizur	Dr. Panjwani Center for Molecular Medicine and Drug Research, International Center for Chemical and Biological Sciences (ICCBS), University of Karachi, Karachi 75270, Pakistan
Rana Waseem Akhtar	Faculty of Veterinary and Animal Sciences, Muhammad Nawaz Shareef University of Agriculture, Multan, Pakistan
Sanaullah Sajid	Institute of Microbiology, University of Agriculture Faisalabad, Pakistan
Shazia Perveen	Department of Zoology, Wildlife and Fisheries University of Agriculture Faisalabad Sub Campus, Toba Tek Singh 36050, Pakistan
Sol Carriazo	Department of Nephrology and Hypertension, IIS-Foundation Jimenaz-Dia--UAM, Madrid, Spain
Si Yuan Khor	Michigan State University, East Lansing, MI, United States
Yasmine Elkeraie	Kidney and Urology Center, Alexandria, Egypt
Yehia Saleh	Department of Cardiology, Houston Methodist DeBakey Heart & Vascular Center, Houston, Texas, USA

CHAPTER 1

Clinical and Diagnostic Implications of Glycated Albumin in Diabetes Mellitus: An Update

Km Neelofar [2,*], **Jamshed Haneef** [1] and **Farah Khan** [2]

[1] *Department of Pharmaceutical Chemistry, School of Pharmaceutical Education and Research, Jamia Hamdard, Hamdard Nagar, New Delhi 110062, India*

[2] *Department of Biochemistry, School of Chemical and Life Sciences, Jamia Hamdard, Hamdard Nagar, New Delhi 110062, India*

Abstract: In diabetes mellitus (DM), non-enzymatic glycation of proteins, lipids, and fatty acids is accelerated due to persistent hyperglycemia and plays an important role in diabetes and its associated secondary complications. Glycation has the potential to alter the biological, structural, and functional properties of macromolecules. Glycated products (early and late) are both involved in provoking the immune-regulatory cells and generating autoantibodies in diabetic patients. More precisely, human serum albumin is the most abundant protein in circulation involved in glycation. Glycated albumin may accumulate in the body tissues of diabetic patients and participate in its secondary complications. This chapter compiles the studies focused on changes in the secondary and tertiary structure of proteins upon glucosylation. Various *in-vitro* and *in-vivo* approaches involved in investigating such changes are systematically reviewed. Besides, the potential role of glycated albumin in the pathogenesis of diabetes mellitus, as well as its applicability as a diagnostic marker in the progression of the disease, is also highlighted.

Keywords: Hyperglycemia, Non-enzymatic glycation, Glycated Albumin, Protein glycation, Diabetes.

INTRODUCTION

Diabetes mellitus (DM) is a metabolic disorder resulting from defects in insulin secretion and/ or action Author, or both. It is characterized by hyperglycemia, polydipsia, glucosuria, and polyuria. In type 1 diabetes, there is a complete absence of insulin, which affects the metabolism of proteins, carbohydrates, and fats. It is a very common autoimmune disease nowadays, afflicting millions of people in India and worldwide also. The disease occurs as a consequence of the organ-specific immune destruction of insulin-producing beta cells within the

* **Corresponding author Km Neelofar:** Department of Biochemistry, School of Chemical and Life Sciences, Jamia Hamdard, Hamdard Nagar, New Delhi 110062, India; E-mail: neloferbiotech@gmail.com

pancreas. However, type 2 diabetes mellitus is the result of the inability of islet beta cells to produce adequate insulin and has become an epidemic. The global prevalence of diabetes in 2011 was 366 million; however, by 2030, it is expected to reach 552 million [1]. Type 2 diabetes mellitus is highly prevalent and accounts for 90–95% of cases. In 21st century, diabetes will be a huge burden due to its increasing global prevalence and higher frequency of chronic complications (nephropathy, retinopathy, neuropathy, and cardiovascular disease), affecting various tissues, difficulty in controlling the disease, and its high cost. During diabetes, persistent hyperglycaemia leads to non-enzymatic glycation of various proteins such as haemoglobin, proteins of the erythrocyte membrane, insulin, human serum albumin (HSA), high and low-density lipoproteins, IgM, IgG, collagen, and histones [2, 3]. Proteins are glycated when glucose is chemically bound to amino groups of proteins without the help of an enzyme, which many structural and conformational changes in protein and proceeds to various micro and macro complications in diabetic patients [4].

Non-enzymatic Glycation

Prof. Louis Camille Maillard gave Millard reaction after his own studies describing the brown colour formed while heating carbohydrate and amine mixtures. It was first described during the early 20th century. Non-enzymatic glycation is a common chemical modification that involves the condensation of a carbohydrate's aldehyde group with either the epsilon group of lysine, hydroxylysine, side chains of arginine, cysteine, and histidine residues [5] or the alpha-amino group of a protein's N terminal amino acid [6]. Only open forms of sugars react with proteins, and a labile aldimine (Schiff base) is formed in a few hours by attaching protein amino group with sugar *via* nucleophilic attack. This product is reversible and can go back to glucose and protein again, or it can form ketoamine, which is slightly reversible. Further, this can undergo intermolecular rearrangement through acid_base catalysis to form 1_amino_1_deoxy fructose (fructosamine), a more stable early glycated product named amadori product in a few days. Both Schiff base and amadori products *in vivo* predominantly exist in the cyclic form [7]. Further, the stable amadori product gradually evolves to a heterogeneous population of fluorescent adducts with new cross-links, which are called advanced glycation end products (AGEs) by irreversible chemical reactions involving oxidation and fragmentation [8] (Fig. **1**). Thus, by subsequent degradation of amadori products and the fragmentation of Schiff base, alpha dicarbonyl compounds and alpha-keto aldehydes formed, respectively (Fig. **2**) [9]. Throughout the 1980s and 1990s, a large body of evidence has implicated that AGEs are mediators of various complications of diabetes and aging. The AGEs also interact with various AGE receptors as RAGEs and stimulate signaling pathways that are important to cause long_term complications in diabetic patients.

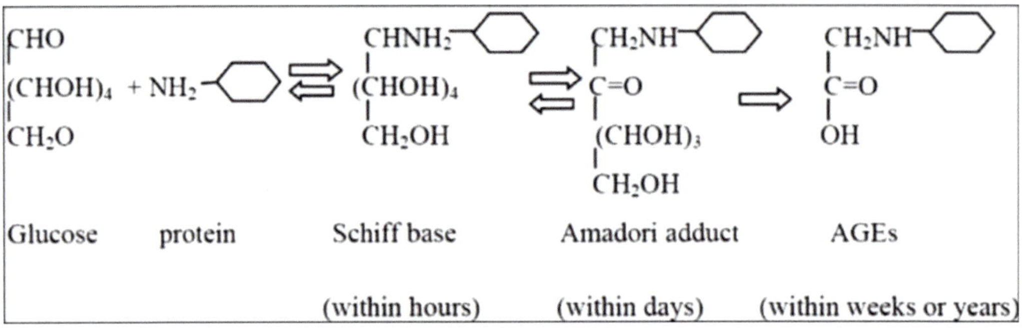

Fig. (1). Non-enzymatic glycation of protein by glucose and production of early and late glycation product. [Source; (Km Neelofar *et al*, 2015).

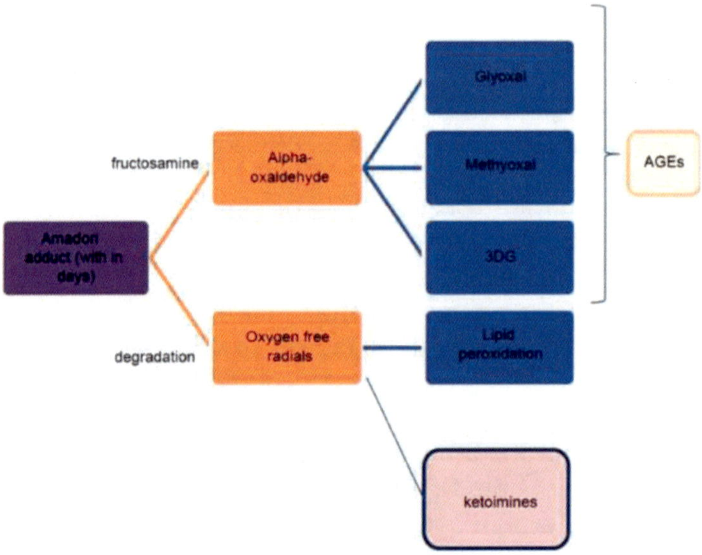

Fig. (2). Amadori adduct fate (Km Neelofar *et al*, 2015).

Non-enzymatic Glycation in Diabetes

Recent studies demonstrate that non-enzymatic glycation is accelerated during hyperglycemia, and its products are aggressively involved in the pathogenesis of diabetes. In diabetes, persistent hyperglycemia leads to non-enzymatic glycation of various proteins such as hemoglobin, proteins of the erythrocyte membrane, insulin, IgG, IgM, human serum albumin, high and low-density lipoproteins, collagen, and histones. Non-enzymatic glycation is also found in normal conditions, but in diabetes, it is increased [10]. Glycated serum proteins consider a marker for hyperglycaemia in diabetes mellitus. Our research studies have shown that early glycation products induced significant changes in albumin structure and function [11]. Glycated proteins are involved in disease pathogenesis by generating free radicals [12]. In prolonged chronic hyperglycemia, the production of free radicals through auto-oxidation of aldehyde group of glucose has enhanced. Non-enzymatic glycation of protein leads to an increase in the flux of glucose through the polyol pathway [13]. Reactive oxygen species (ROS) cause extensive deterioration in protein structure and form neo-epitopes. This new form contributes to its immunogenic potential in diabetic patients and its associated complications [14, 15]. Glucose or small residues are covalently attached with autologous proteins and other biomolecules that can generate conjugates which are efficient to induce an immune response in the host cells. Fabrication of specific glucose-derived adducts on biomacromolecules could function in a manner to form autoantibodies in diabetic patients amadori-albumin is an independent and potent trigger of molecular mediators' contributory to diabetic secondary complications. McCance *et al.* reported an independent association of Amadori adduct with diabetic nephropathy and diabetic retinopathy [16]. Animal studies demonstrated that elevated amadori-albumin promotes a generalized vasculopathy [17] and has been implicated in the development of diabetic nephropathy [18] and retinopathy [19]. Furthermore, amadori-albumin has been reported to be localized in glomeruli of diabetic nephropathy patients [20]. In addition, various intracellular and extracellular glycated proteins have potential roles in diabetes and its complications (Table **1**). Studies have shown that early and advanced glycated adducts have an important role in the development of various diabetic complications such as nephropathy, neuropathy, retinopathy, and cardiovascular diseases.

Table 1. Role of the glycated adduct in diabetes (Km Neelofar *et al.*, 2015).

Early and Advanced Glycated Protein	As a Causative Agent in Diabetes Mellitus
Human serum albumin	Type 1 and type 2 diabetes with retinopathy and nephropathy
Collagen	Diabetic retinopathy
Immunoglobulins (IgG, IgA, IgM)	Type 1 and type 2 diabetes with nephropathy
Plasma Proteins	Type 2 diabetes
Low-density lipoproteins	Diabetic atherosclerosis
Histone	Diabetes

Human Serum Albumin

Albumin is the most abundant and largest protein among all serum proteins in human [21]. HSA is mainly synthesized in the liver and presents 50% of the normal individual's plasma protein with a normal concentration of 30–50 g/l. Albumin plays an important role in physiological, pharmacological, and other functions [22]. It is also involved in the binding and transport capacities of fatty acids, hormones, drugs, and metabolites, the defensive role of oxidative stress, and oncotic pressure regulation. It regulates microvascular permeability and has anti_thrombotic, anti_inflammatory, antioxidation activity. Structurally, albumin is a single-chain globular protein with 585 amino acids involving 1 tryptophan, 1 free cysteine, 59 lysine, and other amino acids. In crystal structure view, HSA looks like a heart-shaped molecule that is divided into three domains [4]. HSA is non enzymatically attached to the glucose molecules and forms glycated-HSA (Fig. **3**). HSA is a lysine-rich protein, and some specific lysine residues are involved in non-enzymatic glycation [9, 11]. Lysine, arginine, and cysteine residues have high nucleophile properties, so they are subjected to glycation mostly.

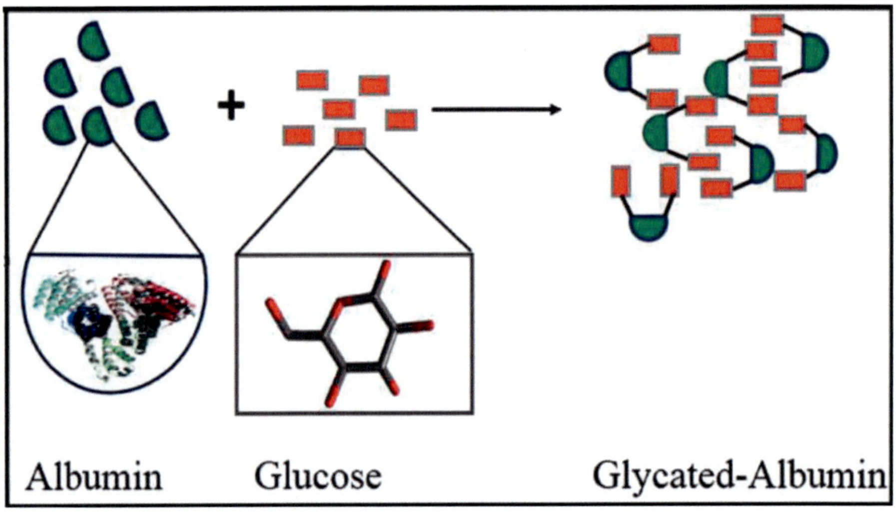

Fig. (3). Albumin binds with glucose form glycated albumin. (Km Neelofar *et al*, 2017).

In HSA, Lysine-525 is considered the prime site for glycation, and it is involved 30% of the overall glycation [15]. For instance, there are other three main sites (Lys-351, Lys-475, and Arg-117) that carbohydrates could bind to HSA. HSA may protect other serum proteins from glycation in the initial stages of diabetes [16]. Arg-410 is also an important site for glycation [17]. Arg-114, Arg-160, Arg-186, Arg-218, and Arg-428 are also involved in glycation [18]. Cys-34 also plays an important role in the glycation process because of its thiol group, which is a powerful nucleophile. Methylglyoxal reacts with this thiol group and forms AGEs such as S-carboxymethyl cysteine (CMC) [19]. *In-vivo* studies have demonstrated that the proportion of glycated albumin in healthy subjects is in the range of 1-10% [20], compared with diabetic individuals [21]. Glycation efficiency depends on the nature and the polymerization of the carbohydrate involved in the process. As an example, ribose induces a faster glycation process with albumin than glucose and forms amyloid-like products [22]. The process of glycation is very important, especially in diabetes and its related complications. It can modify/change the structural and functional properties of intra and extracellular protein and serum proteins. Our studies have shown that early glycation induced significant structural changes in HSA correspond to glucose concentrations upon early glycation [11]. Furthermore, Arif *et al.* reported the highly immunogenic potential of glycated albumin due to the generation of neo_epitopes [23].

Albumin Structure Upon Glycation

Non-enzymatic glycation is one of the underlying modifications that can change protein's primary, secondary and tertiary structure [24, 25]. The glycation of protein induces several structural modifications [26] that can be determine by UV and fluorescence spectroscopy [27 - 30], radiolabelling [31, 32], colorimetric assays [33], circular dichroism [30, 34], and NMR spectroscopy [35]. In addition, other advanced techniques like immunoassays [31, 36], electrophoresis [37], high-performance liquid chromatography (HPLC) [38 - 40], and dynamic light scattering (DLS) [41] have provided information on the total glycation levels or on the number of specific AGEs that are present within albumin. Some of these approaches again include the prior isolation of glycation-induced modified albumin by methods such as boronate affinity chromatography [42 - 44]. More detailed information on glycation-induced modifications has been obtained by using mass spectrometry. Moreover, to locate and identify glycation sites in albumin, liquid chromatography-mass spectrometry (LC-MS) and liquid chromatography-tandem mass spectrometry (LC-MS/MS) have been used [45 - 50].

Secondary structure changes of glycated protein have been detected by Fourier transform infrared spectroscopy (FTIR). Gas chromatography-mass spectrometry (GCMS) has been utilized to investigate glycation at the N-terminus of glycated proteins [51]. To estimate the overall extent of molecular weight by glycation, matrix-assisted laser desorption/ionization time-of-flight mass spectrometry (MALDI-TOF MS) has been used. Neelofar *et al.* reported early glycated product was formed when albumin was incubated with glucose by LCMS using commercial standard furosine [41]. The glycation process may have a variety of physiological effects on protein and other macromolcecules. *In-vitro*, glucose-induced modification in protein is considered as an appropriate model to determine the structural and functional alterations relevant to diabetes mellitus [52]. Structural stability is the most important aspect in carrying out any protein's native functions. Modified protein can be involved in disease progression [53].

Biological Properties of Albumin Upon Glycation

Glycation can modify protein structural properties, and after this modification protein, functional properties may also be changed. This structural and functional modified albumin can be involved in diabetes-associated complications such as retinopathy, neuropathy nephropathy, and coronary artery disease [54]. The deleterious effects of glycated albumin have been highlighted in many research studies. These studies focused on the physiopathological association between glycated albumin and diabetic secondary complications [55]. Research studies have been proven that antioxidant property is strongly affected by non-enzymatic glycation [24, 37, 56]. An amadori-albumin causes lipid peroxidation by generating oxygen free radicals at the potential of Hydrogen 7 [57]. Non-enzymatic glycation of albumin reduces drug binding affinity and transport property [28]. It is reduced by 50% for bilirubin and 20% for long-chain fatty acid as compared to non-glycated HSA. Several *in-vitro* studies suggest that glycated albumin is also involved in platelet activation and aggregation [58, 59]. The pathogenic role of glycated albumin can also be observed in the glucose metabolism of adipocyte cells and skeletal muscle [60]. It has been found that in mouse adipocyte cell lines, glycated albumin triggers the production of intracellular reactive oxygen species that cause inhibition of glucose uptake, resulting in attenuation of adipocyte insulin sensitivity and microangiopathy [61]. Proteins such as Calnexin, a transmembrane protein, and nucleophosmin in monocyte play a role as receptors for early glycated albumin [62]. Amadori-albumin is transported across the renal glomerular capillaries by mesanglial and epithelial cells. This involvement of Amadori-albumin consequent increase in oxidative products that play a strong role in nephropathy development [63]. The role of Amadori-albumin is also reported in diabetic retinopathy [64]. Interaction of glycated albumin with its specific receptor, called RAGEs, affects cellular biology. A signal transduction activates by this interaction and form reactive oxygen species [65]. Cellular oxidative stress activates a cascade of intracellular signals involving MAP-kinase pathways and p21 ras. These pathways phosphorylate extracellular signal-regulated kinase (ERK) [66] and culminate in the activation of the NFkB transcription factor [67 - 73]. Most research studies have shown the role of AGEs in diabetes and associated vascular complications. But now, the role of Amadori proteins in diabetes pathogenesis comes under consideration [74]. Early glycation products contribute to the development of diabetic secondary complications [56, 75]. Amadori albumin, like AGEs, could upregulate the expression of various cellular signaling pathways *via* their receptors to activate NF-kB and AP-1 [69, 71, 76].

Immunological Properties of Albumin Upon Glycation

Early and advanced glycated products of serum proteins show their immunological potential. When the glycated product is injected into the experimental animals, it gives antibodies titre and triggers the immune system (Fig. **4**). Glycation-modified proteins are immunologically active that can induce a substantial immune response. Such glycation-induced modifications may generate the neo-epitopes on the protein surface and become more immunogenic [77]. Many articles have reported that proteins become immunogenic upon glycation. When injected in experimental animals, glycation might change protein conformation results recognized as a foreign particle and give the antibodies titre [78, 79]. Various research reports have documented the presence of autoantibodies in sera of diabetic patients against glycated proteins [80]. Also, the presence of anti-glycated-albumin autoantibodies have been reported in diabetic patients with or without secondary complications [81].

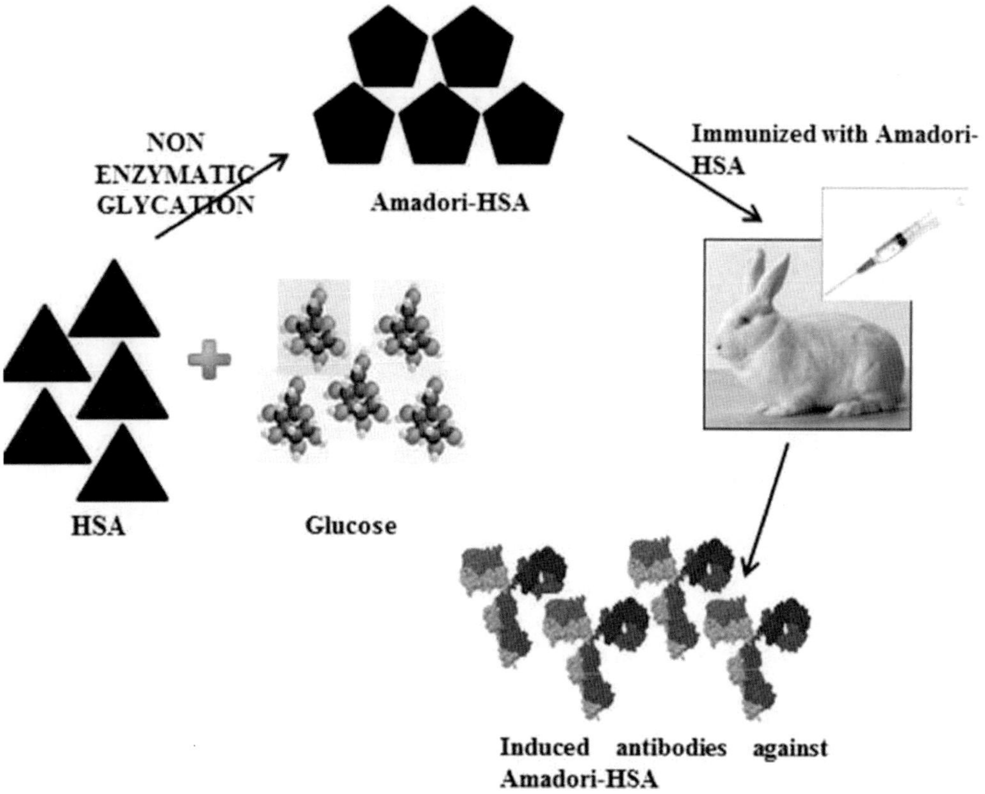

Fig. (4). A Schematic illustration depicting immunogenic potential of Amadori -albumin to induce the generation of antibodies that have specificity for Amadori-albumin. (Km Neelofar *et al.*, 2017).

Glycated proteins worked as an antigen in experimental animals to induce antibodies. These antibodies is highly specific to their corresponding antigen [82]. The binding of induced antibodies against native as well as glycated forms of different proteins is determined by inhibition assay. As a result, these antibodies exhibited a variable degree of recognition for other glycated proteins. Therefore, these outcomes indicate that induced antibodies showed polyspecificity, which means they shared the common epitopes with glycated-albumin and glycated forms of different proteins [8, 83]. Moreover, anti -glycated- protein-IgG antibodies also showed native protein. It showed that all epitopes of native protein had not been changed into neo-epitopes upon glycation. Hence immunization with glycated-albumin may induce polyspecific antibodies, which can detect old and neo-epitopes.

Further, the presence of autoantibodies against glycated proteins were found in the sera of type 1 as well as type 2 diabetic patients with or without secondary complications [84 - 86]. In our previous study, we reported the presence of autoantibodies in diabetic patients' sera with chronic kidney disease against amadori-albumin [41]. Nadeem *et al.* (2013) reported the presence of autoantibodies against glycated-lysine residues in Type1 and Type2 diabetic patients [87]. Turner *et al.* (1997) documented islet cell autoimmunity in type2 diabetic patients with the detection of autoantibodies against glutamic acid decarboxylase (GAD) and islet cytoplasm [88, 89]. The role of these circulating autoantibodies in the pathogenesis of type2 diabetes is less understood. It may adversely disturb intracellular biochemical pathways. The presence of anti-glycated albumin autoantibodies diabetic patients' sera proves that glycated-albumin is immunogenic and can elicit immune response [90, 91]. These autoantibody binds to glycated-albumin and form an immune complex that might be involved in the development of diabetes complications [92]. These complexes may accumulate in the tissues of various organs. In the tissues, they act like pathogenic factors and cause inflammation. However, the level by which inflammation reaction overlaps with autoimmunity is still not known. Now, the research should be focused on autoimmune involvement in type2 diabetes with humoral immune response and chronic inflammation. However, the researcher should focus on detecting the autoantibodies against serum glycated proteins in diabetes and other diseases patients. These autoantibodies can give a great way to diagnosis the disease at an early stage. However, more studies are needed in this direction.

Glycated Albumin as a Diagnosis Marker

Glycated hemoglobin (HbA1c) and blood glucose (BG) levels are the two main clinical parameters to diagnose diabetes [93]. BG level: short-term indicator reflects blood glucose level over a 24 hrs period. Although, HbA1c is known as the gold standard parameter to manage diabetes and its associated secondary complications. But it is also considered as the long-term standard parameter. HbA1c level reflects the glycemic state over the last 2 months due to erythrocytes having a long-term half-life (about 120 days). However, patients with blood-related complications give false HbA1c values. Patients who have iron deficiency, hemolytic anemia, and hemodialysis showed an invalid correlation to BG and HbA1c [94]. Thus, in such cases, hbA1c is not a suitable diagnosis marker as a control [95].

Other types of diagnostic markers like non-protein markers exist, *i.e.*, 1,5-anhydroglucitol (1,5-AG) and self-monitoring of blood glucose levels in the serum. Under the normal condition, glomeruli filtered 1,5-AG from the circulating blood, and renal tubules reabsorbed it completely. Because of the similar structure of 1,5- AG and glucose, they compete for reabsorption. As a result, blood glucose level (180 mg/dl) increases while 1,5 AG level decreases [96]. Though 1,5-AG shows postprandial excursions more correctly than HbA1c and Fructosamine, it does not reflect mean glucose level [97]. It provides information on hyperglycemic excursions [98].

Hemoglobin A1c (HbA1c) and fructosamine (FA) are non-enzymatically glycated proteins that are used to monitor glycemic status in type 2 diabetic patients [99]. They have been commonly used as the primary glycemic control markers, but now glycated albumin (GA) has gained more attention as a new diabetic marker due to some superiority over HbA1c, fructosamine, and other markers. Therefore, to overcome the drawbacks of other markers and achieve a better glycemic status, a novel idea of using glycated albumin as an intermediate glycemic index has been developed (Table **2**).

Table 2. Classification of glycemic markers.

Classification	Glycemic marker	Glycemic status (days)	References
Long term	HbA1c	120	American diabetes association
Intermediate	GA, FA	18-20	Cohen *et al.* (1999)
Short term	1,5-AG and Apo-B	3-5	Buse *et al.* (2003)

GA: Glycated albumin, FA: Fructosamine, HbA1c: Hemoglobin A1c,
1,5-AG: 1,5-anhydroglucitol, Apo-B: Apolipoprotein B

Fructosamine is also similar to glycated albumin, reflecting the glycemic level over 2–3 weeks, but fructosamine refers to all glycated serum proteins, including GA. Like hemoglobin, fructosamine is not influenced by hemoglobin-related disease but is strongly affected by the concentration of proteins in serum and low molecular weight molecules present in plasma-like hemoglobin, bilirubin, and uric acid, *etc.* At the same time, GA is not affected by other proteins concentrations [100, 101]. GA consists shorter half-life of 21 days as compared to haemoglobin. So, it can be considered as a shorter-term glycemic marker as a control for diabetic patients. GA level could not be easily affected by abnormal haemoglobin metabolism [10] and by the lifespan of red blood cells (RBS). The advantage of GA considered as a marker is founded on two facts. First, non-enzymatic glycation of serum albumin is approximately 9 times more than haemoglobin. Secondly, the glycation of albumin occurs ten times more quickly than haemoglobin [32]. All these factors make GA, a good additional diagnostic marker for assessing glycemic control in type1 and type2 diabetes [102]. In many research studies, GA is recommended as an optional marker for glycemic control in hemodialysis patients or gestational diabetes [103] and Alzheimer's disease [104] and diabetes-related complications, including retinopathy [105], nephropathy [106]. It is also documented that glomerular filtration rate (GFR) is negatively connected with HbA1c concentration, and it can change the association of HbA1c with mean glucose, whereas GA values are unaffected. So, GA might be a better glycemic marker for diabetic patients with renal impairment. To detect hyperglycemic status, GA has been shown to be superior to fructosamine alone [107, 108] with the Oral Glucose Tolerance Test (OGTT) as the diagnostic standard [109, 110]. Moreover, from the diagnostic point of view, the GA/HbA1c ratio is very useful for detecting patients with postprandial hyperglycemia or large glycemic excursion [111]. With all these concerns, GA could be used as a short-term glycemic marker as a control. But there are some limitations with GA also. Like, in thyroid dysfunction, nephrotic syndrome, or liver cirrhosis in which the amounts of albumin are affected, glycated albumin level is not a suitable indicator in these cases [112]. Similarly, glycated albumin could be influenced by other conditions, such as body mass index (BMI). Therefore, combined detection of HbA1C and GA may improve the efficacy of diagnosis and improvement of a novel therapeutic potential.

Glycated Albumin Measurements

The American Diabetes Association (ADA) and the European Association Diabetes Study (EASD) recommend "patient-centered" management of glycemic control in patients with Type 2 Diabetes Mellitus (T2DM) and the selection only of biomarkers, such as GA, that reflect the individual health status of the diabetic patient, maintaining the balance between risks and benefits [113]. GA levels are measured as a ratio of total glycated amino acid concentration to albumin concentration. In the old literature, various colorimetric assays were used for the quantification of GA, such as thiobarbituric acid and bromcresol green assays [114]. But these assays have now been replaced by nitroblue tetrazolium (NBT) assay [115] and 2-keto-glucose with hydrazine [116]. Presently GA concentration is also measured with several methods, including ion-exchange chromatography, affinity chromatography and high-performance liquid chromatography (HPLC), immunoassay, enzyme-linked immunosorbent assay (ELISA), enzyme-linked boronate immunoassay, and electrochemical methods.

Recently, an innovative and very promising electrochemical immunoassay has also been developed using nanozymes. This assay shows good linearity and a lower limit of detection [117]. Interestingly, another method that has been recently analyzed is an enzymatic method that shows good analytical performance (Lucic ® GA-L kit, Asahi Kasei Pharma Corporation, Tokyo, Japan). This method is a GA-L kit, Asahi Kasei Pharma Corporation, Tokyo, Japan). This is based on the elimination of endogenous glycated amino acids and peroxides involving the enzymes ketamine oxidase and peroxidase. The glycated albumin is then hydrolyzed by an albumin-specific proteinase and then oxidized by a ketamine oxidase. The hydrogen peroxide produced is then measured quantitatively by the classic colorimetric method of Trinder. Meanwhile, in parallel, the concentration of albumin is measured by the bromocresol violet method, allowing the results to be expressed as the ratio between GA and total albumin [118]. The result of GA is provided as a percentage (GA%) of total albumin. The GA Upper Reference Limit (URL) of 14.5% (95% CI: 14.3–14.7) has been established in Caucasian healthy subjects [5].

Despite the possible benefits of GA, the lack of normal reference data on GA might limit its use as a diagnostic marker for diabetic patients. A study has established that the reference interval of GA in the Japanese population was 12.3–16.9%, and Hiramatsu *et al.* reported (2012) that in healthy Japanese pregnant women, a reference range of GA is 11.5-15.7% [119]. On the other hand, a Chinese research study reported a GA value of 17.1% to be an optimal cut-off in the Chinese population for the diagnosis of diabetes. In the United States, many laboratories used affinity chromatography to state reference values for GA in the range of 0.6–3.0% but by the enzymatic assay GA reference range came out to be 11-16%. It is also needed to establish a GA reference range among the Indian population with or without diabetes before the use of GA as a biomarker for diabetes control.

CONCLUDING REMARKS

This chapter discussed the role of non-enzymatic glycation in the onset and progression of diabetes and its associated complications. The main focus was on the structural and functional properties of glycated albumin. It is interesting that structural and functional impaired glycated albumin becomes more immunogenic and induces an immune response in experimental animals. Further, the presence of autoantibodies against glycated albumin in diabetic patients is also discussed. Glycation-induced modifications in serum proteins can be clinically significant. Different types of biochemical and biophysical techniques have been employed to determine the types of modifications induced in albumin upon glycation. It is more interesting to use glycated albumin as a biomarker to control the blood glucose level over short-to-intermediate periods. Now, various methods are available in clinical laboratories to measure the glycated albumin. To monitor glycemic control in diabetic patients, glycated albumin can be used as a complementary biomarker along with blood glucose and Hb1Ac. However, data on the Indian population are deficient. Further research is warranted to launch the cut-off value in the Indian population using control groups.

CONSENT FOR PUBLICATION

Not applicable.

CONFLICT OF INTEREST

The author declares no conflict of interest, financial or otherwise.

ACKNOWLEDGEMENTS

Declared none.

REFERENCE

[1] Whiting DR, Guariguata L, Weil C, Shaw J. IDF diabetes atlas: global estimates of the prevalence of diabetes for 2011 and 2030. Diabetes Res Clin Pract 2011; 94(3): 311-21. [http://dx.doi.org/10.1016/j.diabres.2011.10.029] [PMID: 22079683]

[2] Ansari NA, Moinuddin , Ali R. Glycated lysine residues: a marker for non-enzymatic protein glycation in age-related diseases. Dis Markers 2011; 30(6): 317-24. [http://dx.doi.org/10.1155/2011/718694] [PMID: 21725160]

[3] Arfat MY, Ashraf JM, Arif Z, Moinuddin , Alam K. Fine characterization of glucosylated human IgG by biochemical and biophysical methods. Int J Biol Macromol 2014; 69: 408-15. [http://dx.doi.org/10.1016/j.ijbiomac.2014.05.069] [PMID: 24953604]

[4] Monnier VM, Sell DR, Dai Z, Nemet I, Collard F, Zhang J. The role of the amadori product in the complications of diabetes. Ann N Y Acad Sci 2008; 1126: 81-8. [http://dx.doi.org/10.1196/annals.1433.052] [PMID: 18448799]

[5] Schalkwijk CG, Miyata T. Early- and advanced non-enzymatic glycation in diabetic vascular

complications: the search for therapeutics. Amino Acids 2012; 42(4): 1193-204. [http://dx.doi.org/10.1007/s00726-010-0779-9] [PMID: 20960212]

[6] Münch G, Schicktanz D, Behme A, *et al.* Amino acid specificity of glycation and protein-AGE crosslinking reactivities determined with a dipeptide SPOT library. Nat Biotechnol 1999; 17(10): 1006-10. [http://dx.doi.org/10.1038/13704] [PMID: 10504703]

[7] John WG, Lamb EJ. The Maillard or browning reaction in diabetes. Eye (Lond) 1993; 7(Pt 2): 230-7. [http://dx.doi.org/10.1038/eye.1993.55] [PMID: 7607341]

[8] Nguyen S, Pascariu M, Ghitescu L. Early glycation products of endothelial plasma membrane proteins in experimental diabetes. Biochim Biophys Acta 2006; 1762(1): 94-102. [http://dx.doi.org/10.1016/j.bbadis.2005.08.004] [PMID: 16139995]

[9] Lee C, Yim MB, Chock PB, Yim HS, Kang SO. Oxidation-reduction properties of methylglyoxal-modified protein in relation to free radical generation. J Biol Chem 1998; 273(39): 25272-8. [http://dx.doi.org/10.1074/jbc.273.39.25272] [PMID: 9737992]

[10] Guthrow CE, Morris MA, Day JF, Thorpe SR, Baynes JW. Enhanced nonenzymatic glucosylation of human serum albumin in diabetes mellitus. Proc Natl Acad Sci USA 1979; 76(9): 4258-61. [http://dx.doi.org/10.1073/pnas.76.9.4258] [PMID: 291961]

[11] Neelofar KM, Ahmad J, Alam K. Impact of *in vitro* nonenzymatic glycation on biophysical and biochemical regimes of human serum albumin: relevance in diabetes associated complications. RSC Advances 2015; 5: 63605-14. [http://dx.doi.org/10.1039/C5RA07232H]

[12] Palm F, Cederberg J, Hansell P, Liss P, Carlsson PO. Reactive oxygen species cause diabetes-induced decrease in renal oxygen tension. Diabetologia 2003; 46(8): 1153-60. [http://dx.doi.org/10.1007/s00125-003-1155-z] [PMID: 12879251]

[13] Giugliano D, Ceriello A, Paolisso G. Oxidative stress and diabetic vascular complications. Diabetes Care 1996; 19(3): 257-67. [http://dx.doi.org/10.2337/diacare.19.3.257] [PMID: 8742574]

[14] Rasheed Z, Khan MWA, Ali R. Hydroxyl radical modification of human serum albumin generated cross reactive antibodies. Autoimmunity 2006; 39(6): 479-88. [http://dx.doi.org/10.1080/08916930600918472] [PMID: 17060027]

[15] Khan MWA, Sherwani S, Khan WA, Ali R. Characterization of hydroxyl radical modified GAD65: a potential autoantigen in type 1 diabetes. Autoimmunity 2009; 42(2): 150-8. [http://dx.doi.org/10.1080/08916930802468276] [PMID: 19085183]

[16] McCance DR, Dyer DG, Dunn JA, *et al.* Maillard reaction products and their relation to complications in insulin-dependent diabetes mellitus. J Clin Invest 1993; 91(6): 2470-8. [http://dx.doi.org/10.1172/JCI116482] [PMID: 8514859]

[17] Cohen MP, Clements RS, Cohen JA, Shearman CW. Glycated albumin promotes a generalized vasculopathy in the db/db mouse. Biochem Biophys Res Commun 1996; 218(1): 72-5. [http://dx.doi.org/10.1006/bbrc.1996.0014] [PMID: 8573179]

[18] Cohen MP, Hud E, Wu VY. Amelioration of diabetic nephropathy by treatment with monoclonal antibodies against glycated albumin. Kidney Int 1994; 45(6): 1673-9. [http://dx.doi.org/10.1038/ki.1994.219] [PMID: 7933815]

[19] Cohen MP, Ziyadeh FN. Role of Amadori-modified nonenzymatically glycated serum proteins in the pathogenesis of diabetic nephropathy. J Am Soc Nephrol 1996; 7(2): 183-90. [http://dx.doi.org/10.1681/ASN.V72183] [PMID: 8785386]

[20] Rondeau P, Navarra G, Militello V, Bourdon E. On the Aggregation of Albumin: Influences of the Protein Glycation. In: Douglas A, Ed. Protein Aggregation. Stein Nova Science Publishers, Inc 2010.

[21] Evans TW. Review article: albumin as a drug--biological effects of albumin unrelated to oncotic pressure. Aliment Pharmacol Ther 2002; 16 (Suppl. 5): 6-11. [http://dx.doi.org/10.1046/j.1365-2036.16.s5.2.x] [PMID: 12423448]

[22] Sakai H, Jinde K, Suzuki D, Yagame M, Nomoto Y. Localization of glycated proteins in the glomeruli of patients with diabetic nephropathy. Nephrol Dial Transplant 1996; 11 (Suppl. 5): 66-71. [http://dx.doi.org/10.1093/ndt/11.supp5.66] [PMID: 9044311]

[23] Arif B, Ashraf JM, Moinuddin , Ahmad J, Arif Z, Alam K. Structural and immunological characterization of Amadori-rich human serum albumin: role in diabetes mellitus. Arch Biochem Biophys 2012; 522(1): 17-25. [http://dx.doi.org/10.1016/j.abb.2012.04.005] [PMID: 22516656]

[24] Shaklai N, Garlick RL, Bunn HF. Nonenzymatic glycosylation of human serum albumin alters its conformation and function. J Biol Chem 1984; 259(6): 3812-7. [http://dx.doi.org/10.1016/S0021-9258(17)43168-1] [PMID: 6706980]

[25] Nakajou K, Watanabe H, Kragh-Hansen U, Maruyama T, Otagiri M. The effect of glycation on the structure, function and biological fate of human serum albumin as revealed by recombinant mutants. Biochim Biophys Acta 2003; 1623(2-3): 88-97. [http://dx.doi.org/10.1016/j.bbagen.2003.08.001] [PMID: 14572906]

[26] Rondeau P, Singh NR, Caillens H, Tallet F, Bourdon E. Oxidative stresses induced by glycoxidized human or bovine serum albumin on human monocytes. Free Radic Biol Med 2008; 45(6): 799-812. [http://dx.doi.org/10.1016/j.freeradbiomed.2008.06.004] [PMID: 18616999]

[27] Arif B, Ashraf JM, Moinuddin A, Ahmad J, Arif Z, Alam K. Structural and immunological characterization of Amadori-rich human serum albumin: role in diabetes mellitus. Arch Biochem Biophys 2012; 522(1): 17-25. [http://dx.doi.org/10.1016/j.abb.2012.04.005] [PMID: 22516656]

[28] Zoellner H, Hou JY, Hochgrebe T, *et al.* Fluorometric and mass spectrometric analysis of nonenzymatic glycosylated albumin. Biochem Biophys Res Commun 2001; 284(1): 83-9. [http://dx.doi.org/10.1006/bbrc.2001.4924] [PMID: 11374874]

[29] Westwood ME, Thornalley PJ. Molecular characteristics of methylglyoxal-modified bovine and human serum albumins. Comparison with glucose-derived advanced glycation endproduct-modified serum albumins. J Protein Chem 1995; 14(5): 359-72. [http://dx.doi.org/10.1007/BF01886793] [PMID: 8590604]

[30] Barzegar A, Moosavi-Movahedi AA, Sattarahmady N, *et al.* Spectroscopic studies of the effects of glycation of human serum albumin on L-Trp binding. Protein Pept Lett 2007; 14(1): 13-8. [http://dx.doi.org/10.2174/092986607779117191] [PMID: 17266645]

[31] Iberg N, Flückiger R. Non-enzymatic glycosylation of albumin *in vivo*. Identification of multiple glycosylated sites. Identification of multiple glycosylated sites. J Biol Chem 1986; 261: 13542-5. [http://dx.doi.org/10.1016/S0021-9258(18)67052-8] [PMID: 3759977]

[32] Day JF, Thorpe SF, Baynes JW. Non-enzymatically glycosylated albumin. J Biol Chem 1979; 254: 595-7. [http://dx.doi.org/10.1016/S0021-9258(17)37845-6] [PMID: 762083]

[33] Baker JR, Zyzak DV, Thorpe SR, Baynes JW. Chemistry of the fructosamine assay: D-glucosone is the product of oxidation of Amadori compounds. Clin Chem 1994; 40(10): 1950-5. [http://dx.doi.org/10.1093/clinchem/40.10.1950] [PMID: 7923778]

[34] Okabe N, Hashizume N. Drug binding properties of glycosylated human serum albumin as measured by fluorescence and circular dichroism. Biol Pharm Bull 1994; 17(1): 16-21. [http://dx.doi.org/10.1248/bpb.17.16] [PMID: 8148809]

[35] Howard MJ, Smales CM. NMR analysis of synthetic human serum albumin α-helix 28 identifies structural distortion upon amadori modification. J Biol Chem 2005; 280(24): 22582-9.

[http://dx.doi.org/10.1074/jbc.M501480200] [PMID: 15837789]

[36] Taneda S, Monnier VM. ELISA of pentosidine, an advanced glycation end product, in biological specimens. Clin Chem 1994; 40(9): 1766-73.
[http://dx.doi.org/10.1093/clinchem/40.9.1766] [PMID: 8070089]

[37] Lee C, Yim MB, Chock PB, Yim HS, Kang SO. Oxidation-reduction properties of methylglyoxal-modified protein in relation to free radical generation. J Biol Chem 1998; 273(39): 25272-8.
[http://dx.doi.org/10.1074/jbc.273.39.25272] [PMID: 9737992]

[38] Schleicher E, Wieland OH. Specific quantitation by HPLC of protein (lysine) bound glucose in human serum albumin and other glycosylated proteins. J Clin Chem Clin Biochem 1981; 19(2): 81-7.
[http://dx.doi.org/10.1515/cclm.1981.19.2.81] [PMID: 6783732]

[39] Ahmed N, Argirov OK, Minhas HS, Cordeiro CAA, Thornalley PJ. Assay of advanced glycation endproducts (AGEs): surveying AGEs by chromatographic assay with derivatization by 6-aminoquinolyl-N-hydroxysuccinimidyl-carbamate and application to Nepsilon-carboxymethyl-lysine- and Nepsilon-(1-carboxyethyl)lysine-modified albumin. Biochem J 2002; 364(Pt 1): 1-14.
[http://dx.doi.org/10.1042/bj3640001] [PMID: 11988070]

[40] Ahmed N, Thornalley PJ. Chromatographic assay of glycation adducts in human serum albumin glycated *in vitro* by derivatization with 6-aminoquinolyl-N-hydroxysuccinimidyl-carbamate and intrinsic fluorescence. Biochem J 2002; 364(Pt 1): 15-24.
[http://dx.doi.org/10.1042/bj3640015] [PMID: 11988071]

[41] Neelofar KM, Ahmad J, Arif Z, Alam K. Elucidating the impact of high concentration of glucose with human serum albumin: a multi technique approach. Int J Biol Macromol 2016; 92: 881-91.
[http://dx.doi.org/10.1016/j.ijbiomac.2016.07.104] [PMID: 27492562]

[42] Lui X, Scouten WH. Handbook of affinity chromatography. In: Hage DS, Ed. Boca Raton, FL: CRC Press 2006; Chap 8.

[43] Zhang Q, Monroe ME, Schepmoes AA, *et al.* Comprehensive identification of glycated peptides and their glycation motifs in plasma and erythrocytes of control and diabetic subjects. J Proteome Res 2011; 10(7): 3076-88.
[http://dx.doi.org/10.1021/pr200040j] [PMID: 21612289]

[44] Frolov A, Hoffmann R. Analysis of amadori peptides enriched by boronic acid affinity chromatography. Ann N Y Acad Sci 2008; 1126: 253-6.
[http://dx.doi.org/10.1196/annals.1433.060] [PMID: 18448825]

[45] Gadgil HS, Bondarenko PV, Treuheit MJ, Ren D. Screening and sequencing of glycated proteins by neutral loss scan LC/MS/MS method. Anal Chem 2007; 79(15): 5991-9.
[http://dx.doi.org/10.1021/ac070619k] [PMID: 17571855]

[46] Kisugi R, Kouzuma T, Yamamoto T, *et al.* Structural and glycation site changes of albumin in diabetic patient with very high glycated albumin. Clin Chim Acta 2007; 382(1-2): 59-64.
[http://dx.doi.org/10.1016/j.cca.2007.04.001] [PMID: 17482589]

[47] Lapolla A, Fedele D, Reitano R, *et al.* Enzymatic digestion and mass spectrometry in the study of advanced glycation end products/peptides. J Am Soc Mass Spectrom 2004; 15(4): 496-509.
[http://dx.doi.org/10.1016/j.jasms.2003.11.014] [PMID: 15047055]

[48] Priego-Capote F, Scherl A, Müller M, Waridel P, Lisacek F, Sanchez JC. Glycation isotopic labeling with 13C-reducing sugars for quantitative analysis of glycated proteins in human plasma. Mol Cell Proteomics 2010; 9(3): 579-92.
[http://dx.doi.org/10.1074/mcp.M900439-MCP200] [PMID: 19955080]

[49] Frost L, Chaudhry M, Bell T, Cohenford M. *in vitro* galactation of human serum albumin: analysis of the protein's galactation sites by mass spectrometry. Anal Biochem 2011; 410(2): 248-56.
[http://dx.doi.org/10.1016/j.ab.2010.11.034] [PMID: 21112314]

[50] Ahmed N, Thornalley PJ. Peptide mapping of human serum albumin modified minimally by

methylglyoxal *in vitro* and *in vivo*. Ann N Y Acad Sci 2005; 1043: 260-6. [http://dx.doi.org/10.1196/annals.1333.031] [PMID: 16037246]

[51] Robb DA, Olufemi OS, Williams DA, Midgley JM. Identification of glycation at the N-terminus of albumin by gas chromatography-mass spectrometry. Biochem J 1989; 261(3): 871-8. [http://dx.doi.org/10.1042/bj2610871] [PMID: 2803249]

[52] Guerin-Dubourg A, Catan A, Bourdon E, Rondeau P. Structural modifications of human albumin in diabetes. Diabete Metab 2012; 38(2): 171-8. [http://dx.doi.org/10.1016/j.diabet.2011.11.002] [PMID: 22349032]

[53] Neelofar K, Arif Z, Alam K, Ahmad J. Hyperglycemia induced structural and functional changes in human serum albumin of diabetic patients: a physico-chemical study. Mol Biosyst 2016; 12(8): 2481-9. [http://dx.doi.org/10.1039/C6MB00324A] [PMID: 27226040]

[54] Brownlee M. The pathological implications of protein glycation. Clin Invest Med 1995; 18(4): 275-81. [PMID: 8549013]

[55] Chen S, Cohen MP, Ziyadeh FN. Amadori-glycated albumin in diabetic nephropathy: pathophysiologic connections. Kidney Int Suppl 2000; 77: S40-4. [http://dx.doi.org/10.1046/j.1523-1755.2000.07707.x] [PMID: 10997689]

[56] Ali O, Bernett G, Comstock P, Estenoz J, Gugliucci A, Roohk HV. Glycated Albumin: an Intermediate Glycemic Index an Analytical Review. Corporate Information Series – 2. Irvine: Epinex Diagnostics Inc. 2008.

[57] Neelofar K, Ahmad J. Amadori albumin in diabetic nephropathy. Indian J Endocrinol Metab 2015; 19(1): 39-46. [http://dx.doi.org/10.4103/2230-8210.146863] [PMID: 25593824]

[58] Cohen MP, Masson N, Hud E, Ziyadeh F, Han DC, Clements RS. Inhibiting albumin glycation ameliorates diabetic nephropathy in the db/db mouse. Exp Nephrol 2000; 8(3): 135-43. [http://dx.doi.org/10.1159/000020661] [PMID: 10810230]

[59] Candiloros H, Muller S, Ziegler O, Donner M, Drouin P. Role of albumin glycation on the erythrocyte aggregation: an *in vitro* study. Diabet Med 1996; 13(7): 646-50. [http://dx.doi.org/10.1002/(SICI)1096-9136(199607)13:7<646::AID-DIA139>3.0.CO;2-J] [PMID: 8840099]

[60] Miele C, Riboulet A, Maitan MA, *et al.* Human glycated albumin affects glucose metabolism in L6 skeletal muscle cells by impairing insulin-induced insulin receptor substrate (IRS) signaling through a protein kinase C alpha-mediated mechanism. J Biol Chem 2003; 278(48): 47376-87. [http://dx.doi.org/10.1074/jbc.M301088200] [PMID: 12970360]

[61] Unoki H, Yamagishi S. Advanced glycation end products and insulin resistance. Curr Pharm Des 2008; 14(10): 987-9. [http://dx.doi.org/10.2174/138161208784139747] [PMID: 18473850]

[62] Wu VY, Shearman CW, Cohen MP. Identification of calnexin as a binding protein for Amadori-modified glycated albumin. Biochem Biophys Res Commun 2001; 284(3): 602-6. [http://dx.doi.org/10.1006/bbrc.2001.4982] [PMID: 11396943]

[63] Ziyadeh FN, Cohen MP. Effects of glycated albumin on mesangial cells: evidence for a role in diabetic nephropathy. Mol Cell Biochem 1993; 125(1): 19-25. [http://dx.doi.org/10.1007/BF00926830] [PMID: 8264568]

[64] Hasan NA. Effects of trace elements on albumin and lipoprotein glycation in diabetic retinopathy. Saudi Med J 2009; 30(10): 1263-71. [PMID: 19838431]

[65] Wautier JL, Zoukourian C, Chappey O, *et al.* Receptor-mediated endothelial cell dysfunction in diabetic vasculopathy. Soluble receptor for advanced glycation end products blocks hyperpermeability

in diabetic rats. J Clin Invest 1996; 97(1): 238-43.
[http://dx.doi.org/10.1172/JCI118397] [PMID: 8550841]

[66] Cohen MP, Shea E, Chen S, Shearman CW. Glycated albumin increases oxidative stress, activates NF-kappa B and extracellular signal-regulated kinase (ERK), and stimulates ERK-dependent transforming growth factor-beta 1 production in macrophage RAW cells. J Lab Clin Med 2003; 141(4): 242-9.
[http://dx.doi.org/10.1067/mlc.2003.27] [PMID: 12677169]

[67] Lander HM, Tauras JM, Ogiste JS, Hori O, Moss RA, Schmidt AM. Activation of the receptor for advanced glycation end products triggers a p21(ras)-dependent mitogen-activated protein kinase pathway regulated by oxidant stress. J Biol Chem 1997; 272(28): 17810-4.
[http://dx.doi.org/10.1074/jbc.272.28.17810] [PMID: 9211935]

[68] Yan SD, Schmidt AM, Anderson GM, *et al.* Enhanced cellular oxidant stress by the interaction of advanced glycation end products with their receptors/binding proteins. J Biol Chem 1994; 269(13): 9889-97.
[http://dx.doi.org/10.1016/S0021-9258(17)36966-1] [PMID: 8144582]

[69] Hattori Y, Kakishita H, Akimoto K, Matsumura M, Kasai K. Glycated serum albumin-induced vascular smooth muscle cell proliferation through activation of the mitogen-activated protein kinase/extracellular signal-regulated kinase pathway by protein kinase C. Biochem Biophys Res Commun 2001; 281(4): 891-6.
[http://dx.doi.org/10.1006/bbrc.2001.4436] [PMID: 11237743]

[70] Hattori Y, Suzuki M, Hattori S, Kasai K. Vascular smooth muscle cell activation by glycated albumin (Amadori adducts). Hypertension 2002; 39(1): 22-8.
[http://dx.doi.org/10.1161/hy1201.097300] [PMID: 11799073]

[71] Fan X, Subramaniam R, Weiss MF, Monnier VM. Methylglyoxal-bovine serum albumin stimulates tumor necrosis factor alpha secretion in RAW 264.7 cells through activation of mitogen-activating protein kinase, nuclear factor kappaB and intracellular reactive oxygen species formation. Arch Biochem Biophys 2003; 409(2): 274-86.
[http://dx.doi.org/10.1016/S0003-9861(02)00599-4] [PMID: 12504894]

[72] Naitoh T, Kitahara M, Tsuruzoe N. Tumor necrosis factor-alpha is induced through phorbol ester--and glycated human albumin-dependent pathway in THP-1 cells. Cell Signal 2001; 13(5): 331-4.
[http://dx.doi.org/10.1016/S0898-6568(01)00152-8] [PMID: 11369514]

[73] Abordo EA, Thornalley PJ. Synthesis and secretion of tumour necrosis factor-alpha by human monocytic THP-1 cells and chemotaxis induced by human serum albumin derivatives modified with methylglyoxal and glucose-derived advanced glycation endproducts. Immunol Lett 1997; 58(3): 139-47.
[http://dx.doi.org/10.1016/S0165-2478(97)00080-1] [PMID: 9293394]

[74] Schalkwijk CG, Lieuw-a-Fa M, van Hinsbergh VW, Stehouwer CD. Pathophysiological role of Amadori-glycated proteins in diabetic microangiopathy. Semin Vasc Med 2002; 2(2): 191-7.
[http://dx.doi.org/10.1055/s-2002-32042] [PMID: 16222610]

[75] Brandt R, Landmesser C, Vogt L, *et al.* Differential expression of fructosyllysine-specific receptors on monocytes and macrophages and possible pathophysiological significance. Diabetologia 1996; 39(10): 1140-7.
[http://dx.doi.org/10.1007/BF02658499] [PMID: 8897000]

[76] Higai K, Shimamura A, Matsumoto K. Amadori-modified glycated albumin predominantly induces E-selectin expression on human umbilical vein endothelial cells through NADPH oxidase activation. Clin Chim Acta 2006; 367(1-2): 137-43.
[http://dx.doi.org/10.1016/j.cca.2005.12.008] [PMID: 16455066]

[77] Mackay IR. Clustering and commonalities among autoimmune diseases. J Autoimmun 2009; 33(3-4): 170-7.

[http://dx.doi.org/10.1016/j.jaut.2009.09.006] [PMID: 19837564]

[78] Wentworth JM, Fourlanos S, Harrison LC. Reappraising the stereotypes of diabetes in the modern diabetogenic environment. Nat Rev Endocrinol 2009; 5(9): 483-9. [http://dx.doi.org/10.1038/nrendo.2009.149] [PMID: 19636326]

[79] Jaleel A, Halvatsiotis P, Williamson B, Juhasz P, Martin S, Nair KS. Identification of Amadori-modified plasma proteins in type 2 diabetes and the effect of short-term intensive insulin treatment. Diabetes Care 2005; 28(3): 645-52. [http://dx.doi.org/10.2337/diacare.28.3.645] [PMID: 15735202]

[80] Cohen MP, Ziyadeh FN. Amadori glucose adducts modulate mesangial cell growth and collagen gene expression. Kidney Int 1994; 45(2): 475-84. [http://dx.doi.org/10.1038/ki.1994.62] [PMID: 8164436]

[81] Lapolla A, Traldi P, Fedele D. Importance of measuring products of non-enzymatic glycation of proteins. Clin Biochem 2005; 38(2): 103-15. [http://dx.doi.org/10.1016/j.clinbiochem.2004.09.007] [PMID: 15642271]

[82] Khan MWA, Qadrie ZL, Khan WA. Antibodies against gluco-oxidatively modified human serum albumin detected in diabetes-associated complications. Int Arch Allergy Immunol 2010; 153(2): 207-14. [http://dx.doi.org/10.1159/000312639] [PMID: 20413989]

[83] Sakai H, Jinde K, Suzuki D, Yagame M, Nomoto Y. Localization of glycated proteins in the glomeruli of patients with diabetic nephropathy. Nephrol Dial Transplant 1996; 11 (Suppl. 5): 66-71. [http://dx.doi.org/10.1093/ndt/11.supp5.66] [PMID: 9044311]

[84] Stern MP. Kelly West Lecture. Primary prevention of type II diabetes mellitus. Diabetes Care 1991; 14(5): 399-410. [http://dx.doi.org/10.2337/diacare.14.5.399] [PMID: 2060452]

[85] Ansari NA, Dash D. Amadori glycated proteins: role in production of autoantibodies in diabetes mellitus and effect of inhibitors on non-enzymatic glycation. Aging Dis 2013; 4(1): 50-6. [PMID: 23423609]

[86] Turner R, Stratton I, Horton V, *et al.* UKPDS 25: autoantibodies to islet-cell cytoplasm and glutamic acid decarboxylase for prediction of insulin requirement in type 2 diabetes. Lancet 1997; 350(9087): 1288-93. [http://dx.doi.org/10.1016/S0140-6736(97)03062-6] [PMID: 9357409]

[87] Irvine WJ, McCallum CJ, Gray RS, Duncan LJ. Clinical and pathogenic significance of pancreatic-islet-cell antibodies in diabetics treated with oral hypoglycaemic agents. Lancet 1977; 1(8020): 1025-7. [http://dx.doi.org/10.1016/S0140-6736(77)91258-2] [PMID: 67485]

[88] Tuomi T, Groop LC, Zimmet PZ, Rowley MJ, Knowles W, Mackay IR. Antibodies to glutamic acid decarboxylase reveal latent autoimmune diabetes mellitus in adults with a non-insulin-dependent onset of disease. Diabetes 1993; 42(2): 359-62. [http://dx.doi.org/10.2337/diab.42.2.359] [PMID: 8425674]

[89] Turk Z, Ljubic S, Turk N, Benko B. Detection of autoantibodies against advanced glycation endproducts and AGE-immune complexes in serum of patients with diabetes mellitus. Clin Chim Acta 2001; 303(1-2): 105-15. [http://dx.doi.org/10.1016/S0009-8981(00)00389-2] [PMID: 11163030]

[90] Roohk HV, Zaidi AR. A review of glycated albumin as an intermediate glycation index for controlling diabetes. J Diabetes Sci Technol 2008; 2(6): 1114-21. [http://dx.doi.org/10.1177/193229680800200620] [PMID: 19885300]

[91] Hashimoto K, Osugi T, Noguchi S, *et al.* A1C but not serum glycated albumin is elevated because of iron deficiency in late pregnancy in diabetic women. Diabetes Care 2010; 33(3): 509-11.

[http://dx.doi.org/10.2337/dc09-1954] [PMID: 20032277]

[92] Bry L, Chen PC, Sacks DB. Effects of hemoglobin variants and chemically modified derivatives on assays for glycohemoglobin. Clin Chem 2001; 47(2): 153-63. [http://dx.doi.org/10.1093/clinchem/47.2.153] [PMID: 11159762]

[93] Buse JB, Freeman JL, Edelman SV, Jovanovic L, McGill JB. Serum 1,5-anhydroglucitol (GlycoMark): a short-term glycemic marker. Diabetes Technol Ther 2003; 5(3): 355-63. [http://dx.doi.org/10.1089/152091503765691839] [PMID: 12828817]

[94] Dungan KM, Buse JB, Largay J, *et al.* 1,5-anhydroglucitol and postprandial hyperglycemia as measured by continuous glucose monitoring system in moderately controlled patients with diabetes. Diabetes Care 2006; 29(6): 1214-9. [http://dx.doi.org/10.2337/dc06-1910] [PMID: 16731998]

[95] True MW. Circulating biomarkers of glycemia in diabetes management and implications for personalized medicine. J Diabetes Sci Technol 2009; 3(4): 743-7. [http://dx.doi.org/10.1177/193229680900300421] [PMID: 20144323]

[96] Lee EY, Lee BW, Kim D, *et al.* Glycated albumin is a useful glycation index for monitoring fluctuating and poorly controlled type 2 diabetic patients. Acta Diabetol 2011; 48(2): 167-72. [http://dx.doi.org/10.1007/s00592-010-0242-0] [PMID: 21153482]

[97] Armbruster DA. Fructosamine: structure, analysis, and clinical usefulness. Clin Chem 1987; 33(12): 2153-63. [http://dx.doi.org/10.1093/clinchem/33.12.2153] [PMID: 3319287]

[98] Koga M, Kasayama S. Clinical impact of glycated albumin as another glycemic control marker. Endocr J 2010; 57(9): 751-62. [http://dx.doi.org/10.1507/endocrj.K10E-138] [PMID: 20724796]

[99] Yoshiuchi K, Matsuhisa M, Katakami N, *et al.* Glycated albumin is a better indicator for glucose excursion than glycated hemoglobin in type 1 and type 2 diabetes. Endocr J 2008; 55(3): 503-7. [http://dx.doi.org/10.1507/endocrj.K07E-089] [PMID: 18445997]

[100] Gugliucci A. Glycation as the glucose link to diabetic complications. J Am Osteopath Assoc 2000; 100(10): 621-34. [PMID: 11105451]

[101] Okumura A, Mitamura Y, Namekata K, Nakamura K, Harada C, Harada T. Glycated albumin induces activation of activator protein-1 in retinal glial cells. Jpn J Ophthalmol 2007; 51(3): 236-7. [http://dx.doi.org/10.1007/s10384-007-0431-8] [PMID: 17554491]

[102] Shuvaev VV, Laffont I, Serot JM, Fujii J, Taniguchi N, Siest G. Increased protein glycation in cerebrospinal fluid of Alzheimer's disease. Neurobiol Aging 2001; 22(3): 397-402. [http://dx.doi.org/10.1016/S0197-4580(00)00253-0] [PMID: 11378244]

[103] Nagayama H, Inaba M, Okabe R, *et al.* Glycated albumin as an improved indicator of glycemic control in hemodialysis patients with type 2 diabetes based on fasting plasma glucose and oral glucose tolerance test. Biomed Pharmacother 2009; 63(3): 236-40. [http://dx.doi.org/10.1016/j.biopha.2008.04.002] [PMID: 18538530]

[104] Hay-Lombardie A, Kamel S, Bigot-Corbel E. [Insights on glycated albumin]. Ann Biol Clin (Paris) 2019; 77(4): 407-14. [PMID: 31418702]

[105] Chan CL, Pyle L, Kelsey M, Newnes L, Zeitler PS, Nadeau KJ. Screening for type 2 diabetes and prediabetes in obese youth: evaluating alternate markers of glycemia - 1,5-anhydroglucitol, fructosamine, and glycated albumin. Pediatr Diabetes 2016; 17(3): 206-11. [http://dx.doi.org/10.1111/pedi.12258] [PMID: 25652226]

[106] Chume FC, Kieling MH, Correa Freitas PA, Cavagnolli G, Camargo JL. Glycated albumin as a diagnostic tool in diabetes: An alternative or an additional test? PLoS One 2019; 14(12): e0227065.

[http://dx.doi.org/10.1371/journal.pone.0227065] [PMID: 31891628]

[107] Sacks DB. A1C *versus* glucose testing: a comparison. Diabetes Care 2011; 34(2): 518-23. [http://dx.doi.org/10.2337/dc10-1546] [PMID: 21270207]

[108] Mo Y, Ma X, Li H, *et al.* Relationship between glycated albumin and glycated hemoglobin according to glucose tolerance status: A multicenter study. Diabetes Res Clin Pract 2016; 115: 17-23. [http://dx.doi.org/10.1016/j.diabres.2016.03.003] [PMID: 27242118]

[109] Feskens E, Brennan L, Dussort P, *et al.* Potential Markers of Dietary Glycemic Exposures for Sustained Dietary Interventions in Populations without Diabetes. Adv Nutr 2020; 11(5): 1221-36. [http://dx.doi.org/10.1093/advances/nmaa058] [PMID: 32449931]

[110] Nishimura R, Kanda A, Sano H, *et al.* Glycated albumin is low in obese, non-diabetic children. Diabetes Res Clin Pract 2006; 71(3): 334-8. [http://dx.doi.org/10.1016/j.diabres.2005.07.008] [PMID: 16154660]

[111] *Standards of Medical Care in Diabetes-2019* Abridged for Primary Care Providers. Clin Diabetes 2019; 37(1): 11-34. [http://dx.doi.org/10.2337/cd18-0105] [PMID: 30705493]

[112] Johnson RN, Baker JR. Inaccuracy in measuring glycated albumin concentration by thiobarbituric acid colorimetry and by boronate chromatography. Clin Chem 1988; 34(7): 1456-9. [http://dx.doi.org/10.1093/clinchem/34.7.1456] [PMID: 3390917]

[113] Kumar A, Rao P, Pattabiraman TN. A colorimetric method for the estimation of serum glycated proteins based on differential reduction of free and bound glucose by sodium borohydride. Biochem Med Metab Biol 1988; 39(3): 296-304. [http://dx.doi.org/10.1016/0885-4505(88)90089-8] [PMID: 3395510]

[114] Mashiba S, Uchida K, Okuda S, Tomita S. Measurement of glycated albumin by the nitroblue tetrazolium colorimetric method. Clin Chim Acta 1992; 212(1-2): 3-15. [http://dx.doi.org/10.1016/0009-8981(92)90133-B] [PMID: 1486680]

[115] Choi H, Son SE, Hur W, *et al.* Electrochemical Immunoassay for Determination of Glycated Albumin using Nanozymes. Sci Rep 2020; 10(1): 9513. [http://dx.doi.org/10.1038/s41598-020-66446-3] [PMID: 32528061]

[116] Kohzuma T, Koga M. Lucica GA-L glycated albumin assay kit: a new diagnostic test for diabetes mellitus. Mol Diagn Ther 2010; 14(1): 49-51. [http://dx.doi.org/10.1007/BF03256353] [PMID: 20121290]

[117] Chen L, Zhang B, Yang L, Lou J, Jiang Y, Zhang S. Individualized Correction of the Interference of Hemolysis on Glycated Albumin Determined by the Ketamine Oxidase Method. Lab Med 2020; 51(2): 151-6. [PMID: 31352488]

[118] Hiramatsu Y, Shimizu I, Omori Y, Nakabayashi M. Determination of reference intervals of glycated albumin and hemoglobin A1c in healthy pregnant Japanese women and analysis of their time courses and influencing factors during pregnancy. Endocr J 2012; 59(2): 145-51. [http://dx.doi.org/10.1507/endocrj.K10E-410] [PMID: 22166921]

[119] Ma XJ, Pan JM, Bao YQ, *et al.* Combined assessment of glycated albumin and fasting plasma glucose improves the detection of diabetes in Chinese subjects. Clin Exp Pharmacol Physiol 2010; 37(10): 974-9. [http://dx.doi.org/10.1111/j.1440-1681.2010.05417.x] [PMID: 20557319]

CHAPTER 2

Current Strategies of New Drugs for Diabetes Management

Maliha Sarfraz[1,*], **Rahman M. Hafizur**[2], **Hayat Ullah**[3,*], **Sanaullah Sajid**[4], **Rana Waseem Akhtar**[5], **Mamoona Noreen**[1], **Shazia Perveen**[1] and **Misbah Ullah Khan**[6]

[1] *Department of Zoology, Wildlife and Fisheries University of Agriculture Faisalabad Sub Campus, Toba Tek Singh 36050, Pakistan*

[2] *Dr. Panjwani Center for Molecular Medicine and Drug Research, International Center for Chemical and Biological Sciences (ICCBS), University of Karachi, Karachi 75270, Pakistan*

[3] *Department of Chemistry, University of Okara, Okara 56300, Punjab, Pakistan*

[4] *Institute of Microbiology, University of Agriculture Faisalabad, Pakistan*

[5] *Faculty of Veterinary and Animal Sciences, Muhammad Nawaz Shareef University of Agriculture, Multan, Pakistan*

[6] *Center for Nano-Sciences, University of Okara, Okara 56300, Punjab, Pakistan*

Abstract: Several aspects need to be explored in drug therapy for diabetes patients. Some specific glucose-reducing medicines are present, while other medicines are associated with unintentional changes in hyperglycemia. Diabetes is a developing epidemic that has caused significant socioeconomic problems in several countries throughout the world. Despite scientific discoveries, greater healthcare services, and higher literacy rates, the disease continues to plague many industries, particularly developing countries. The current trends show an increase in premature mortality, which threatens world prosperity. Experimental and technical improvements have been made in sulphonylureas, alpha-glucosidase inhibitors, biguanides, and thiazolidinediones, all of which are beneficial in lowering glucose levels. The latest drug research techniques have led to the development of novel therapeutic groups such as amylin analogs, incretin mimetics, GIP analogs, active peroxisome proliferator receptors, and dipeptidyl peptidase-4 inhibitors as targets for future diabetes therapy medications. Furthermore, drug development and detection for diabetes treatment have been revolutionized by identifying and investigating bioactive compounds from herbs. This chapter discusses vital fields of clinical diabetology regarding opportunities for stem cells and nanotechnology as next-generation therapies, with an emphasis on evolving developments and reviews why plant-derived products are reliably common for treating and managing diabetes.

* **Corresponding authors Maliha Sarfraz and Hayat Ullah:** Department of zoology wildlife and fisheries university of agriculture Faisalabad sub campus Toba tek Singh 36050, Pakistan;
E-mails: maliha.sarfraz@yahoo.com, hayatullah@uo.edu.pk

Shazia Anjum (Ed.)

Keywords: Diabetes, Emerging Trends, Herbal Formulations, Glucose-lowering Drugs.

INTRODUCTION

Diabetes mellitus (DM) is a complicated metabolic condition identical to elevated blood glucose levels or hyperglycemia, resulting from insulin secretion deficiencies, intervention, or both, as displayed in Fig. (**1**). The persistent metabolic disproportion related to this condition places the patient at increased danger of long-standing macro and microvascular problems, leading to repeated hospitalization and complications, including an elevated danger of cardiovascular disease, unless high-quality treatment is provided [1].

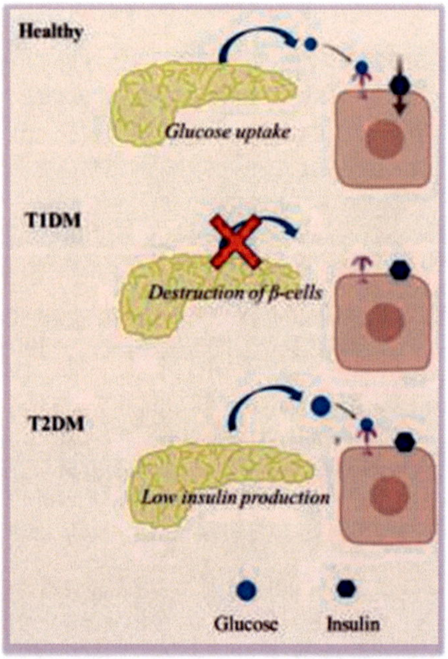

Fig. (1). Diabetes mellitus and its types.

Diabetes is a common and significant global public health concern. According to the International Association of Diabetes (IDF), around 463 million adult diabetes patients were documented worldwide in 2019, which is around 9.3 percent of adults aged 20-79 years, and the number of diabetes patients is still rising [2]. The selection and implementation of glucose control therapy rely on a variety of factors, such as the condition of hyperglycemia, the underlying liver and kidney functions, hypoglycemic risk, the body mass index, capacity to regulate blood glucose, and drug cost. Type 2 diabetes therapeutics include stimulus for insulin release by GLP analogs such as liraglutide and exenatide [3, 4], insulin injection to balance β-cell defects, inhibition of dipeptidyl peptidase-4 (DPP-4) by sitagliptin, and improved islets survival [5, 6] and islet cell regeneration through islet neogenesis associated protein (INGAP) peptide therapy aiming at islet cell regeneration [7].

Diabetes has become a threat to people's health and is a significant global problem for health and society. A timely clinical concern is diabetes care. Along with diet variety and appropriate workouts, antidiabetic medications are important approaches in the treatment of diabetes. Several hypoglycemic agents, like insulin and insulin analogs, biguanides, sulfonylureas, thiazolidinediones, glinides, alpha-glucosidase inhibitors, dipeptidyl peptidase 4 (DPP4), glucagon-like peptide 1 (GLP-1) receptor agonists, and sodium-glucose cotransporter 2 (SGLT2) inhibitors, are currently used in the treatment of diabetes [8]. However, almost half of patients with diabetes cannot achieve treatment goals, including glycemic control, even over 10 years [9 - 11]. A big confusion about the suitable choice and screening of antidiabetic drugs because of the various hypoglycemic drugs is the accessibility and the possibility that the same hypoglycemic agent may contribute to different beneficial responses in each individual. The American Diabetes Association (ADA) and the European Association for the Study of Diabetes (EASD) also propose individualized diabetes attention and precision medicine applications [12, 13]. Providing medication that relates to the genetic knowledge of individuals through pharmacogenomics is one way to achieve precision medicine and direct the proper use of antidiabetic agents [14, 15].

Strategies for the treatment of pharmacologic agents (leptin, β-3-agonists) can increase the resistance of glucose uptake by effectively reducing visceral fat. A function for macrophage fatty-inhibitors (thiazolidinediones, CCR2 antagonists) in treating insulin resistance and vascular disease is also strengthened in different reported studies. Thus, two research lines worth exploring include (i) the interpretation of the visceral fat secretory biology to determine key mediators of the Mets and (ii) drug production for modulative delivery of body fat [16]. Some new kinds of hypoglycemic medicines, such as GLP-1, DPP-IV inhibitors, amylin inhibitors, peroxisome proliferators, and activated receivers, have also been developed and recorded. Any active molecules and bioactive compounds purified from herbs and seeds add to the war on diabetes. These plant components have overturned the production of medicines and led to the discovery of diabetes drugs. Several recent studies have been conducted on important fields of diabetes, focusing more on the statin-based method of diabetes treatment and next-generation antidiabetic stem cell therapy [17].

Also, potential novel diabetes medications have begun to emerge; new targets are projected to provide more reliable care for diabetes. This article would also outline the hypoglycemic drugs and nearly future new goals for diabetes patients' care and discuss some additional successful diabetes therapies. The rising trend in diabetes occurrence and prevalence is alarming and puts a heavy burden on medical expenditure and our modern healthcare system.

CURRENT TREATMENTS FOR TYPE-2 DIABETES MELLITUS

T2D can not be healed permanently, but medications, herbs, and dietary changes can be used to control the severity and symptoms. Drugs of various types, such as biguanides (metformin), sulfonylureas (glyburide and glipizide), meglitinides (repaglinide and nateglinides), and thiazolidinediones, are some of the most commonly employed pharmacological agents for the treatment of T2D. Pioglitazone is the first line of protection, the medications belonging to these groups are prescribed to avoid deterioration of the diabetic disease (Fig. **2**).

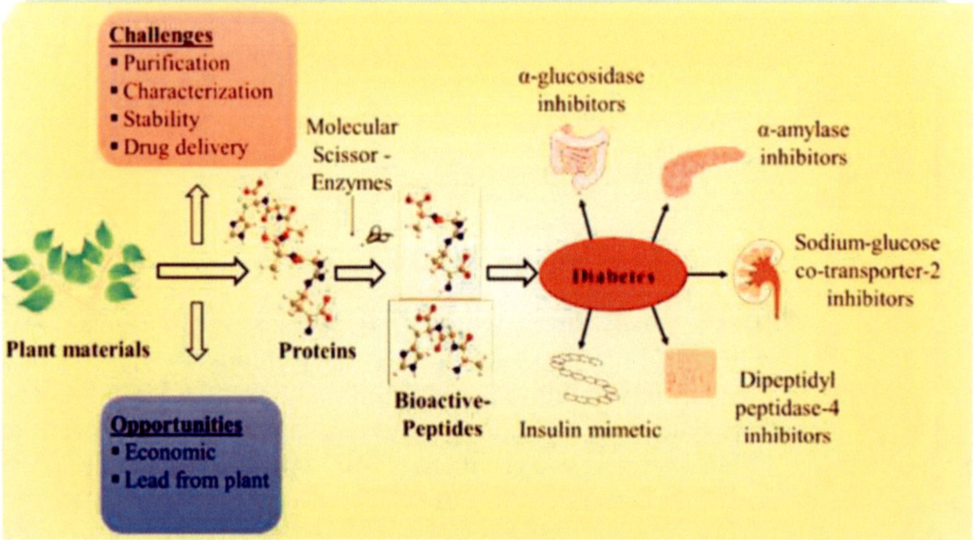

Fig. (2). Common therapeutic approaches for type-2 diabetes mellitus.

Thiazolidinediones

Thiazolidinediones (TZDs) are insulin sensitizers that primarily operate on target organs such as the liver and muscles by improving insulin sensitivity. The mechanism by which TZDs exert their anti-diabetic effect includes stimulation of the peroxisome proliferator-activated receptor (PPAR γ) transcription factor. This element modifies the transcription, namely fatty acyl-CoA synthase, glucokinase, malic enzyme, and glucose transporter 4 (GLUT4), of many genes intricated in lipid metabolism and glucose for energy balance. In this way, insulin tolerance in adipose tissue, liver, and muscle is decreased by TZDs (Fig. **3**) [18]. The increased proliferation of peripheral adipocytes to increase the absorption of free fatty acids is one of the key side effects of PPAR γ receptor activation. This effect can adversely result in weight gain and increased mass of peripheral fat [19]. The possible effect of TZD on coronary events in diabetes patients has been revealed in many recent research and reviews. In this sense, meta-analyses of adverse effect evidence from randomized clinical trials have found that thiazolidinedione usage can be linked with increased myocardial ischemic risk events in the diabetic patient role [20]. Fluid accumulation is another detrimental effect associated with the use of TZDs. It has been indicated that the stimulation of sodium-coupled bicarbonate absorption from the renal proximal tubule may result in the development of TZD-induced edema in the kidney. The increase in sodium and fluid absorption from the renal tubule contributes to an increase in the volume of the kidneys [21]. The findings of this research have sparked controversy and instilled doubt about the application of TZDs in care plans for diabetes.

Biguanide

Biguanides are another insulin sensitizer, and metformin belongs to this family of widely used anti-diabetic medicines. The hypoglycemic activity of metformin, by its action on insulin receptors and glucose transporters presents on target cells, such as the skeletal muscle and liver cells, involves an increase in the use of glucose [22]. It is understood that metformin inhibits the activity of pyruvate dehydrogenase and thus contributes to lactic acidosis, an uncommon but possibly lethal complication linked with metformin usage shown in Fig. (**3**). The elevated risk of lactic acidosis caused by metformin normally occurs in patients with renal, pulmonary, or cardiac insufficiency or a history of liver failure [23].

Sulfonylureas

Sulfonylureas are secretagogues and are meant to enhance the release of insulin from pancreatic beta-cells. The ATP-sensitive potassium (KATP) channels, which regulate the pancreatic β-cell membrane potential, are the primary targets of drugs belonging to the sulfonylurea class. The binding of the drug to the subunit of the sulfonylurea receptor (SUR) of the KATP channel allows the cell membrane to depolarize, leading to calcium ion influx. This results in insulin granule exocytosis from the pancreatic β-cell [24]. For those people who have been taking sulfonylureas (like glyburide) for a longer time, diabetic hypoglycemia is a big concern. Elderly patients and patients who also skip meals are more vulnerable to the risk of sulfonylurea-associated hypoglycemia. Increased cardiovascular risk is also associated with sulfonylurea. Studies suggest that, while inducing the closing of the pancreas β-cell KATP channels to promote insulin secretion, this treatment can also contribute to the closure of the myocardial KATP channels, contributing to a higher frequency of coronary activity in these patients [25]. Sulfonylureas are usually an alternative for hyperglycemia following a biguanide treatment or patients with metformin intolerance. Hypoglycemia is a known side effect of the drug type, and the blood sugar level of patients should be monitored regularly while on a sulphonylurea regimen. These patients should also monitor their body weight and renal function regularly [26].

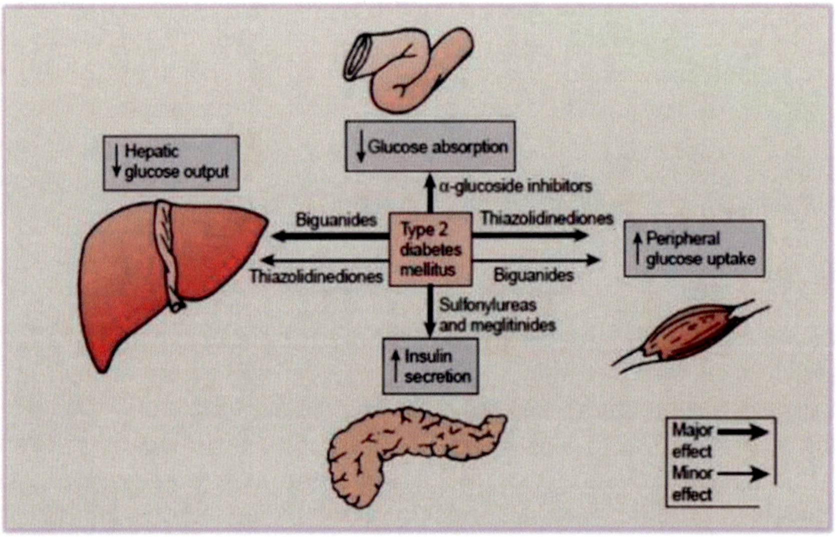

Fig. (3). Site of action of Biguanides, Thiazolidinediones, Sulfonylureas, and alpha glucoside inhibitors.

Meglitinides

The mode of action of meglitinides is like that of sulfonylureas; but, when contrasted with sulfonylureas, its action is regulated by a separate binding site on the SUR of the β-cell [27]. To treat the different co-morbidities involved with T2D, a variety of other medications are also prescribed, along with the anti-diabetic drugs discussed above. Some anti-diabetic drugs are vulnerable to drug-drug reactions, resulting in adverse events and side effects that lead to the diabetic patient's complications. For instance, a known structural analog of sulphonylurea, sulphonamides, replaces it with plasma protein and makes it more freely available for its activity. This potentiates the risk of sulfonylurea-caused hypoglycemia [28]. Some sulfonylureas are metabolized by liver metabolic enzymes, and thus inducers of hepatic drug metabolism, such as rifampicin, increase the clearance of sulfonylureas and hence decrease the plasma concentrations of sulfonylurea and its efficacy [26]. For the optimal care of T2D patients, knowledge of the benefits, as well as the dangers of the large variety of medications available today, is therefore important.

SGLT2 Inhibitors

SGLT2 inhibitors are a new class of drugs for the treatment of type-2 diabetes that act by inhibiting renal glucose reabsorption. They have been adopted rapidly into clinical practice guidelines due to a combination of glucose-lowering with weight reduction [29 - 31]. The difficulty in matching the normal physiology of insulin secretion with exogenous insulin use is a constant challenge when managing type-1 diabetes. As weight gain is a frequent consequence of intensive glucose control [32], the use of adjunctive treatments that promote weight loss is sometimes considered. In addition, there is preliminary evidence that SGLT2 inhibitors may attenuate the progression of kidney disease in type-1 diabetes by decreasing glomerular hyperfiltration [33, 34]. As a result of these preliminary data, some clinicians have considered SGLT2 inhibitors to represent an attractive option in type-1 diabetes, resulting in off-label use in this population. Case reports of diabetic ketoacidosis (DKA) associated with SGLT2 inhibitors started to appear earlier this year [35 - 37]. The US Food and Drug Administration (FDA) published a formal warning regarding this potential complication in May 2015, reporting cases in type-2 diabetes [38]. This has been followed by a similar warning from the European Medicines Evaluation Agency (EMEA) and the manufacturers of the three currently approved SGLT2 inhibitors, dapagliflozin, canagliflozin, and empagliflozin [39] (Fig. **4**).

Insulin

Insulin is a hormone produced by pancreatic beta cells, regulates sugar levels in the bloodstream of the body, and allows excess glucose to be stored mainly in the liver. Although T2D patients produce their insulin, either the quantity of insulin produced is insufficient, or there is a further low insulin response in these patients' target cells as the disease progresses shown in Fig. (**5**). Hence, in most cases, insulin therapy is chosen as the final step for glucose-lowering therapies [40]. The major disadvantages linked with long-term insulin therapy usage are hypoglycemia and weight gain. These symptoms are well-justified because there is a decrease in glycosuria and decreases in energy consumption in these patients as this therapy improves the glycemic level [41]. Additionally, its indirect effect on cell proliferation is a serious problem known to be associated with hyperinsulinemic hypoglycemia. Specific cell proliferation and survival pathways may be improved by hyperinsulinemia (due to its growth-promoting properties), resulting in the risk of cancer progression in various organs, such as the liver, pancreas, colon, and many others. Also, there is the formation of fatty lumps over the sites of injections called lipohypertrophy when insulin is injected into the subcutaneous layer. It is most common in people who receive multiple daily injections frequently, which may affect insulin absorption, leading to changes in blood glucose levels [42 - 44].

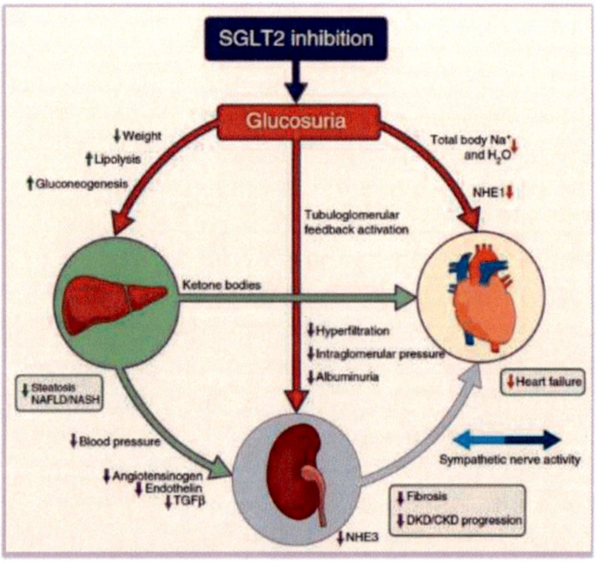

Fig. (4). SGLT 2 inhibitors targeted organs and mechanism of action.

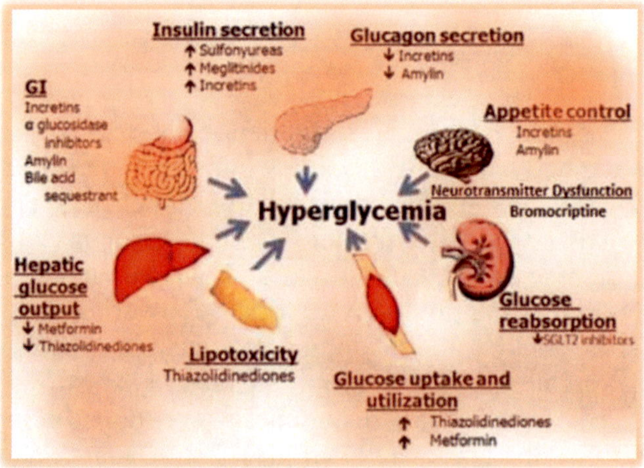

Fig. (5). Insulin and other antidiabetic drugs targeted organs.

Incretin Mimetics

The incretin effect differs from the oral glucose load in the insulin secretory response in contrast to intravenous glucose administration. The incretin effect after oral glucose intake causes 50-70 percent secretion of total insulin [45]. A glucose-dependent insulinotropic polypeptide (GIP, or incretin) and glucagon-like peptide (GLP-1) are the naturally occurring incretin hormone involved in glycemic control; they have a short half-life and are hydrolyzed by DPP-4 inhibitors within 1.5 min. The incretin effect in T2DM patients is less or absent.

While T2DM patients, GIP insulinotropic action is lost. Incretins minimize gastric emptying, resulting in weight decrease and become an important therapeutic method for T2DM treatment. GLP-1 receptor agonists and DPP-4 inhibitors are included in these two drug classes. Clinical data have shown that in patients with T2DM, it causes glycemic control while body weight and blood pressure decreased. Furthermore, hypoglycemia is low (except when used in combination with a sulfonylurea) because of their glucose-dependent mechanism of action [46].

COMPLEMENTARY TREATMENTS FOR THE MANAGEMENT OF T2D

There are high risks related to the use of conventional anti-diabetic agents. They may prove toxic in some cases and may adversely affect the patient's health. Several studies have suggested that lifestyle interventions based on improving physical activity and nutrition may help to better manage the disease as an approach to combating this disease and improving the quality of life for diabetic patients.

It is a well-known reality that physical exercise increases the general quality of life and is likely to avoid multiple lifestyle-related diseases such as cardiovascular disease, obesity, and T2D. The skeletal muscles increase their glucose uptake by many folds over a daily stretch of physical activity, thereby reducing hyperglycemic conditions in the blood [47]. Physical exercise speed and duration are the two main factors that decide the type of fuel used for exercise. As muscle glycogen is steadily reduced, there is a change in the supply of energy to circulating glucose, free fatty acids, and greater oxidation of carbohydrates.

The origin of circulating glucose also shifts to gluconeogenesis from hepatic glycogenolysis [48]. A meta-analysis of 8538 patients showed that more than 150 minutes of structured exercise training, including aerobic exercise, resistance training, or maybe both, cause a decrease in HbA1c as compared to 150 minutes or less per week [49]. Similarly, a systemic review of 10 prospective cohort studies shows that moderately severe physical activity, such as walking, is also linked with decreased risk of T2D [50].

Different clinical findings in diabetic patients indicate a drop in HbA1c with the help of aerobic workouts, resistance, stretching, upper and lower body. Boule *et al.* have examined the effects of multiple exercises over 8 weeks to research their effect on HbA1C and body mass in 504 T2D patients and analyzed them using mathematical models provided by the above study [48]. Also, the research conducted by Ishii *et al.*, with the help of appropriate exercises, promotes an increase in insulin sensitivity [51]. Cuff *et al.* observed 28 postmenopausal T2D patients subjected to aerobic strength training for 16 weeks and the findings were assessed by hyperinsulinemic-euglycemic clamp glucose disposal and computed tomography scans of abdominal and mid-thigh skeletal muscles, resulting in an increase in infusion rates and a reduction in exercise community muscle density compared with control g Other experiments undertaken by Castanenda *et al.*, Dustan *et al.*, incomparable lines add more proof [52 - 55].

NATURAL PRODUCTS WITH ANTI-DIABETIC PROPERTIES

To cure diabetes and its accompanying diseases, a growing number of herbs are utilized. The latest NAPRALERT database lists more than 1300 plant species spanning more than 750 genera in 190 families, including nearly all higher plant forms with lower plants, like fungi and algae. In conventional medicine, many herbs are utilized as antidiabetics, especially for T2DM [56, 57]. A total of 21,000 plants, used for medical purposes worldwide, of which over 400 for diabetes care are available, have been identified by the World Health Organization (WHO). While several herbal medications for the treatment of diabetes are available, only a limited number of those plants have undergone scientific and medical assessments to evaluate their effectiveness. Any of the antidiabetic medicinal plants used are trigonella foenum-graecum, Allium sativum, Caesalpinia bondu, and Ferrulaassafoetida. The antidiabetic aspect of medicinal plants is responsible for the existence of phenolic compounds, flavonoids, terpenoids, and coumarins. The blood glucose levels were lowered by these components. Any examples of branded medicines derived from natural sources and used as antidiabetic drugs include picalnogenol, acarbosis, miglitol, and voglibose [58]. A few studies described the actual anti-diabetic action, but several herbs have been experimentally considered to confirm their physiological activity. However, many chemical constituents are identified and isolated, like alkaloids, carbohydrates, glycopeptides, peptides, terpenoids, amines, steroids, lipids, coumarins, flavonoids, sulphur compounds, and inorganic ions [59]. Few examples of herbs that are utilized for diabetes therapies are *Gymnema sylvestre*, *Momordica charantia*, *Trigonella foenum-graceum*, *Azadirachta indica*, *Curcuma longa,*

Piper nigrum, and *Phoenix dactylifera* [60 - 64]. The suggested mechanisms of action of these herbs are they directly release an insulin secretion, and hepatic glycolysis regulation, glycogenesis, adrenomimeticism, the potassium channel blocker activity of pancreatic beta cells, stimulation of cAMP, and control of intestinal glucose absorption [65 - 67] shown in Fig. (**6**).

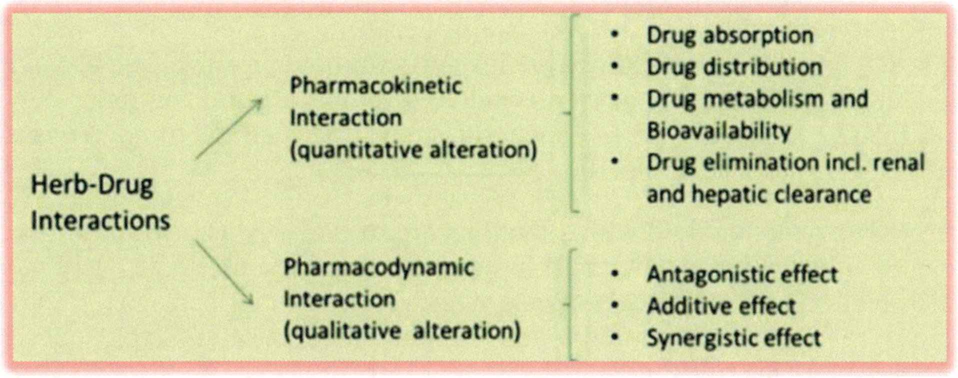

Fig. (6). Mechanism of action of herbal drugs.

It is suggested that cinnamon has many health advantages, such as the capacity to regulate blood glucose, overall amounts of cholesterol and triglycerides, etc. The active ingredient in cinnamon, cinnamonaldehyde, contributes primarily to the promotion of insulin secretion and glucose uptake [68]. It improves glucose uptake by inducing insulin receptor kinase activity, thereby contributing to auto-phosphorylation of the insulin receptors, which in turn stimulates pathway cascade, which eventually results in activation of GLUT4 [69]. In Pakistan, the first clinical trial to research the role of cinnamon in regulating T2D was performed on 60 diabetic individuals who were supplemented with different doses of cinnamon, and it was observed that the mean fasting serum glucose, triglycerides, low lipid density, cholesterol, and total cholesterol levels were substantially reduced after 40 days relative to placebo groups who did not consult [70]. A meta-analysis of 10 RCTs with a sample size of 543 patients found a substantial impact of cinnamon on blood glucose and concluded that intake of cinnamon was correlated with statistically significantly lower glucose, low-density lipid cholesterol (LDL-C), total cholesterol, and triglyceride concentrations [68]. Thus, it may be speculated from multiple types of research that cinnamon use in the diet could help cure diabetes.

It has been used in cooking all over the world to impart taste and fragrance. In addition to this, for various medicinal reasons, its antioxidant function has made it useful. Studies have shown that an antioxidant isolated from garlic, S-allyl cysteine sulfoxide, can lead to its beneficial impact on diabetes [71]. The clinical trial performed in a population of 50 T2D patients with hyperlipidemia found that intake of 900mg/day garlic powder tablets for 6 weeks dramatically lowered overall cholesterol, LDL-C, systolic blood pressure, and improved lipid-cholesterol (HDL-C) high-density [72]. Another randomized, single-blind, placebo-controlled trial, which was performed on 70 T2D patients with newly diagnosed dyslipidemia for 12 weeks, further demonstrated its anti-diabetic efficacy and found that garlic displayed a short-term lipid profile advantage and a greater reduction in total cholesterol and LDL-C and a mild improvement in HDL-C relative to placebo [73]. Some more studies are required to further assess the anti-diabetic properties of garlic.

Berberine is a plant alkaloid with a long history of both Ayurvedic and Chinese therapeutic usage. With medicine. It has a broad spectrum of effects, especially antimicrobial (against bacterial) effects. Diarrhea, intestinal parasites, Candida albicans, yeast, fungal infections, and possibly methicillinin Staphylococcus aureus, resistant) and anti-inflammatory responses. In the roots, it can be found, Rhizomes and stem bark of many species, such as Coptischinensis, Hydrastiscanadensis, Berberisaquifolium, Berberis vulgaris, and Berberisaristata [74], respectively. Berberine, although its mode of action is not well elucidated, has been shown to have anti-diabetic properties. One of the pathways indicating the health benefits of berberine is its activity on adenosine monophosphate-activated protein kinase (AMPK), which contributes to the phosphorylation of essential targets such as lipid metabolism enzymes, lipolysis, oxidation of fatty acids, and glucose absorption. Experiments performed in rat models have shown that berberine-induced AMPK activation induces a cascade of events leading to muscle GLUT4 translocation and adipocyte lowering of lipids [75]. There are few case controls studies that provide evidence of Berberine's hypoglycemic effect. In China, an RCT of 106 T2D patients still suffering from dyslipidemia reported a substantial decrease in fasting and postprandial plasma glucose and HbA1 levels when participants received a daily dosage of 1 gm of berberine for 3 months [76]. A meta-analysis of 14 RCTs involving 1068 participants revealed that berberine has beneficial effects on the regulation of blood glucose in T2D patients and has an effect like that of traditional oral hypoglycemics (metformin, glipizide and, rosiglitazone). Furthermore, no significant adverse effects of berberine were found in this study [77]. Long-term studies are possible, however, with larger sample sizes. To better recognize, as an anti-diabetic agent, the mechanism, effectiveness, and protection of berberine.

CURRENT AND FUTURE THERAPIES FOR TYPE 1 DIABETES

A century ago, the discovery of insulin revolutionized the treatment of this lifelong autoimmune condition as well as prolonging the life expectancy of individuals with type 1 diabetes. People with diabetes type 1 continue as an essential therapeutic alternative based on exogenous insulins [78]. The overarching objective of Type 1 immune therapy is to inhibit or postpone the depletion of beta-cell functional mass. Autoimmunity in type 1 diabetes has traditionally been understood to focus on systemic immune dysregulation and autoreactive T cells, which avoid thymic selection and migrate to the outskirts where they kill islets. The outlook on type 1 diabetes pathogenesis was called the "homicide" of the T cell mediation [79]. Many immune-modulatory therapies rely on T-effector with the conventional immune-centric view of Type 1 pathogenesis. Teplizumab and otelixizumab anti-CD3 antibodies have demonstrated a certain attenuation of beta cell loss [80]. Clinical studies have substantially restored C-peptide secretion and better glycemic function for children and adults with new-income type 1 diabetes in the case of low doses of anti-thymocyte globulin (ATG) therapy (*versus* placebo) [81]. The core proinflammatory cytokine TNF-α blockage or antagonistic with infliximab, adalimumab, or receptor-fusion protein etanercept has shown some potential for diabetes type 1, with indications for enhanced C-peptide regulation and secretion [82]. As an appealing target in type 1 diabetes, IL-21 was proposed [83].

Amylin, which is a neuroendocrine hormone, stimulates the release of glucagon, which helps reduce postprandial glucose variability in non-immunomodulatory treatments for type 1 diabetes. The injectable amylin analog pramlintide is only approved in the USA to treat type 1 and type 2 diabetes [84] as an addition to mealtime insulin. SGLT inhibitors lower levels of blood glucose by restricting the absorption of glucose in the small bowel and encouraging kidney excretion [85]. Dapagliflozin, empagliflozin and sotagliflozin results showed that SGLT inhibition was beneficial when insulin was applied to the treatment of type1 diabetes [86]. Phase II results in adults with type 1 diabetes have recently been negative for short-acting GLP-1 RA exenatide. The use of GLP-1 in type 1 diabetes was accompanied by increased rates of symptomatic hypoglycemia and hyperglycemia with ketosis, thereby limiting clinical use in this population [87]. Verapamil is a popular blocker used as an anti-hypertensive for decades. Verapamil has encouraged survival of functional beta cells in mouse models of type 1 diabetes through a pathway that includes decreased expression of the thioredoxin-interacting protein cellular redox regulator [88]. Verapamil was stronger than placebo in a smaller Phase II study in adults with type 1 diabetes with meal-induced C-peptide secretion and no safety risks were found [89].

STEM CELL THERAPEUTIC APPROACH

Different emerging experimental fields of study have ultimately addressed the curiosity of discovering a potential therapeutic for diabetes, with stem cell science being one of them. When the pancreatic beta cells produce insufficient insulin that leads to type 1 and type 2 diabetes. So those therapies that improve cell response to insulin action or help to improve beta-cell defects are preferable. A novel source is the β-cell replacement method through conventional islet cell and pancreatic transplantation strategies are constrained due to donor organ shortages [90] as shown in Fig. (**7**). Unlike autoimmune pancreatic β-cell death cause type 1 diabetes, while type 2 diabetes due to abnormal function of β-cells along with insulin resistance in peripheral organs [91]. Due to its immunosuppressive nature, mesenchymal stem cell (MSC) therapy has appeared as a potential therapy for type 1 diabetes. Because of the direct interaction and development of soluble markers, MSCs have been shown to exhibit immunomodulatory properties in both conditions [92 - 95].

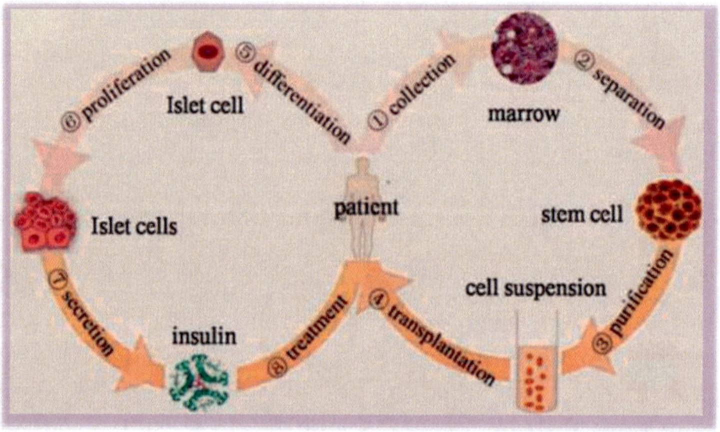

Fig. (7). Steps to produce insulin by stem cell and their transplantation.

MSCs can discriminate into various lineages of mesenchymal cells. Multipotent hematopoietic stem cells can produce all types of cells. This therapy results in increased β-cell activity in newly detected patients with type 1 diabetes [96]. Additional studies have shown that type 1 diabetic patients can generate induced pluripotent stem (iPS) cells by reprogramming three transcription factors (OCT4, SOX2, and KLF4) by their adult fibroblasts. This type of cell known as pluripotent stem cells tempted by diabetes (DiPS) is pluripotent and produces insulin. Type 1 disease modeling and cell replacement therapy; this is beneficial [97]. Some experiments have shown that MSCs originating from the bone marrow can differentiate both *in vitro* and *in vivo* [98 - 100] into insulin-generating cells.

Owing to their pluripotent nature the importance of human embryonic stem cells

(ESCs) for diabetes treatment has drawn excessive interest as shown in Fig. (**8**). The study has many drawbacks, as there is a shortage of effective methods for producing specific types of cells, immunological rejection of transplanted cells, and difficulties in purifying precise lineages [89]. Other issues contain the unchecked production of transplanted embryonic stem cells into a particular type [101].

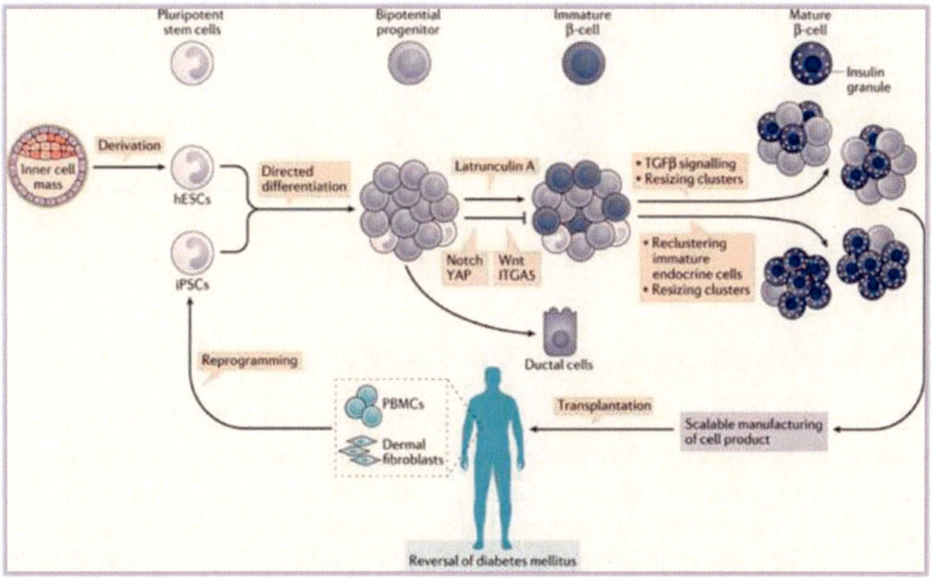

Fig. (8). Production of beta cells by stem cell and reversal of diabetes by transplantation.

NANOTECHNOLOGY AND DIABETES

Novel methods for measuring glucose and the distribution of insulin have been implemented by the nanotechnology interface in the treatment of diabetes. The glucose sensors benefits, and closed-loop insulin therapy methods have been shown by experts in encouraging the treatment of diabetes and make it beneficial [102] for both type 1 and type 2 diabetes. A microcapsule containing pores is a nanomedical system that has become a hopeful instrument for the drug delivery approach. The pores are significantly wide which enables minor molecules like glucose, oxygen, and insulin to pass through, yet they are small to encourage larger molecules of the immune system, like immunoglobulins and graft-borne virusparticles, to travel. Comprising microcapsules Langerhans cell replacement islets, often originating from pigs, may be inserted underneath the skin of patients with diabetes. Without the need for effective immunosuppressants, this could briefly preserve the body's fragile glucose regulation feedback loop, which can put the patient at significant risk of infection [103].

The approach to drug delivery targeted by the nanoparticle has tremendous advantages, including increased drug bioavailability by targeting individual tissues, muscles, and tumors, delivering the maximum drug at the targeted location. The scalability of a nanoparticle is one of the greatest scientific problems. A dynamic activity is the creation of three-dimensional nanostructures relative to stand-alone nanosurfaces. Since processing methods have yet to be consistent. One more fear is that nanoparticle exposure can be harmful or poisonous. There is a growing question about the possible harmful effects of engineered nanomaterials such as carbon buckyballs and nanotubes through inhalation, ingestion, or absorption through the skin [104]. Insulin forms an important requirement for advanced type 1 and type 2 diabetes, and infections, unpleasant administration, and inadequate patient compliance have been included with conventional insulin delivery systems. However, by controlling the delivery of insulin constituting pulmonary, transdermal, nasal, and closed-loop delivery, recent micro-and nanotechnologies have enabled the process of insulin administration [103].

EMERGING TECHNOLOGIES FOR DIABETES TREATMENT

As discussed above, new technology for the delivery of insulin will greatly increase patients' support of intensive care, glycemic management, and life quality of diabetes, though slowing and reducing the risk of complications. Intelligent systems of insulin delivery that can respond to physiological signals or external stimuli to achieve regulated insulin release are suited to physiological conditions. GRIDSs can improve compliance successfully for diabetic patients. For diabetes care, dual and multi-responsive mechanisms that are susceptible to stimulus variation have demonstrated tremendous therapeutic effectiveness as shown in Fig. (**9**). Furthermore, stem cell transplant therapy of diabetes has an effect that is impossible to do by conventional treatments and can minimize complications. It is possible to achieve actual control of blood glucose and diabetes self-management by the Internet and telephone-based techniques. There is also a long way to go until they are commonly used in the pharmacy, considering the impressive successes of these new insulin delivery approaches in diabetes care. Researching and designing a new kind of insulin delivery device with excellent efficiency is a demanding task.

Some scientific questions need to be discussed. First, since medication for diabetes is long-term, severe complications will be caused by the potential side effects. As a result, to plan a highly successful stimulus-sensitive insulin delivery system, material selectivity, durability, biocompatibility, biodegradability, cytotoxicity, and responsive speed should be considered carefully. Second, in the treatment of diabetes with regenerative medicine, it is also important to closely analyze the selection and collection of stem cell types. Finally, the infrastructure focused on the App (Internet and Mobile Phone) lacks quality assurance and control of the information given as shown in Fig. (**9**). Looking ahead to the future, if current findings can be outlined in a timely way and extended to future studies, a range of new diabetes prevention approaches and innovations can undoubtedly be more explored and alternative uses of reasonable clinical drugs can be shown. For myself, the future diabetes treatment is to achieve remote control of smartphone applications. A small blood glucose meter is held by the patient, which constantly tracks the fluctuations in blood glucose in the patient's body for 24 hours and communicates the results through a mobile phone. The handset then monitors the insulin controller inserted in the patient remotely and activates the resulting insulin dose [104].

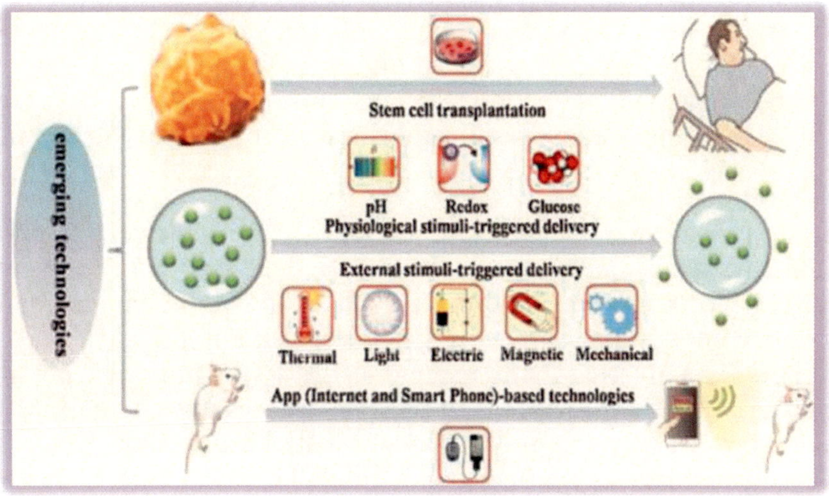

Fig. (9). Schematic presentation of emerging technologies for diabetes treatment.

CONCLUSION

Well-monitored glycemic regulation is required for the treatment of type 2 diabetes. As it may lead to a lack of glycemic regulation, the need to control the gradual degradation of β-cell function is necessary. Conventional medications and insulin are currently in use; nevertheless, the resulting metabolic and glucoregulatory dysfunctions cannot be reversed. The danger of diabetes is increasing day by day. Furthermore, incretin-based therapies and peptide analogs are intense and based on combinational therapy. They can restore and sustain the functioning of β-cells and stop the development of type 2 diabetes. The efficacy and effectiveness of the new medication will depend on its potential to treat/relieve one or more metabolic disorders in the current period, whether it is increased insulin output or increased glucose uptake and use of peripheral tissues, particularly skeletal muscle. A variety of other groups, in addition to new generations of therapeutics, have also been identified as potential methods alone or in conjunction to provide successful diabetes care.

FUTURE PERSPECTIVES

Diabetes, which accounts for a worldwide presence, has stayed the most daunting health issue in the 21st century. Diabetes remains a major public health problem, but the positive news is that substantial strides have been made in diabetes prevention, diagnosis, and care. Patients need insulin administration 3-4 times a day during their lives for the treatment of type 1 diabetes, and their blood sugar levels should be routinely checked for complications such as retinopathy, and the risk of coronary problems needs to be avoided. About 1,300 patients with type 1 diabetes are expected to undergo entire organ (pancreatic) transplantation and do not need insulin injection, but the need for organ transplantation is larger than the availability. Rejection of the transplanted organ is another risk factor; thus, powerful immunosuppressive medications are prescribed for the recipient, which can lead to other severe illnesses.

One of the new developments in the treatment of diabetes is the possibility of leptin therapy. It is an adipocyte-secreted hormone that works on neurons within the central nervous system. The various acts of this hormone include regulating excessive weight gain by suppressing food consumption and increasing energy expenditure. By activating leptin receptors (LEPRs), leptins also control glucose homeostasis. The central nervous system has been shown to control the sugar-lowering activity of leptins; it was thought that neurons in the brain concerning type 1 diabetes may have affected the antidiabetic function of leptins. Leptin therapy through CNS-dependent pathways in mice enhances insulin-deficient type 1 diabetes. Another field of drug study concerns the creation and use of multi-drug mucoadhesive microcapsules such as glipizide to ensure the controlled release and successful targeting of the drug. Mucoadhesion has become a novel method in the design of drug delivery because it triggers the gradual release of the drug at the site of action or absorption, thereby increasing the drug's contact with the underlying tissue, thus, improving the drug's bioavailability. The drug delivery methods that have been followed as a potential solution for diabetes have no end. The approach to transdermal insulin administration (developed because of unpleasant and complex insulin therapy) retains excessive amounts of insulin without insulin deposits in the skin, often with subcutaneous injections of insulin.

Tremendous scientific advancements have been made to completely cure the conditions, but no therapeutic approach has been fully effective. The quest for an appropriate drug is not far ahead, with emerging technology revolutionizing the care possibilities. Extensive experiments have revolutionized the identification of the pathway genes that contribute to the progression of the disease and entire genomes sequencing. Faulty gene identification in organism genome detection has been made with the development of techniques, like PCRs, DNA microarrays, and gene knockouts with silencing. The rising diabetes prevalence worldwide is causing a financial strain on the country's budget. Diabetes treatment is possible, unlike certain other disorders, and if properly controlled, it is very successful in minimizing risks, like heart problems, blindness, amputations, and renal failure. With extensive research, it is not difficult to obtain the best therapeutics for the treatment of diabetes.

CONSENT FOR PUBLICATION

Not applicable.

CONFLICT OF INTEREST

The author declares no conflict of interest, financial or otherwise.

ACKNOWLEDGEMENTS

Declared none.

REFERENCES

[1] Type 2 Diabetes in Adults: Management. London: National Institute for Health and Care Excellence 2015; p. 28.

[2] Buse JB, Rosenstock J, Sesti G, *et al.* Liraglutide once a day *versus* exenatide twice a day for type 2 diabetes: a 26-week randomised, parallel-group, multinational, open-label trial (LEAD-6). Lancet 2009; 374(9683): 39-47.
[http://dx.doi.org/10.1016/S0140-6736(09)60659-0] [PMID: 19515413]

[3] Buse JB, Rosenstock J, Sesti G, *et al.* Liraglutide once a day *versus* exenatide twice a day for type 2 diabetes: a 26-week randomised, parallel-group, multinational, open-label trial (LEAD-6). Lancet 2009; 374(9683): 39-47.
[http://dx.doi.org/10.1016/S0140-6736(09)60659-0] [PMID: 19515413]

[4] Croom KF, McCormack PL. Liraglutide: a review of its use in type 2 diabetes mellitus. Drugs 2009; 69(14): 1985-2004.
[http://dx.doi.org/10.2165/11201060-000000000-00000] [PMID: 19747013]

[5] Kim SJ, Nian C, Doudet DJ, McIntosh CHS. Inhibition of dipeptidyl peptidase IV with sitagliptin (MK0431) prolongs islet graft survival in streptozotocin-induced diabetic mice. Diabetes 2008; 57(5): 1331-9.
[http://dx.doi.org/10.2337/db07-1639] [PMID: 18299314]

[6] Kim SJ, Nian C, Doudet DJ, McIntosh CHS. Dipeptidyl peptidase IV inhibition with MK0431 improves islet graft survival in diabetic NOD mice partially *via* T-cell modulation. Diabetes 2009;

58(3): 641-51.
[http://dx.doi.org/10.2337/db08-1101] [PMID: 19073764]

[7] Rafaeloff R, Pittenger GL, Barlow SW, *et al.* Cloning and sequencing of the pancreatic islet neogenesis associated protein (INGAP) gene and its expression in islet neogenesis in hamsters. J Clin Invest 1997; 99(9): 2100-9.
[http://dx.doi.org/10.1172/JCI119383] [PMID: 9151782]

[8] American Diabetes A. Pharmacologic approaches to glycemic treatment: standards of medical care in diabetes—2020. Diabetes Care 2020; 43: 98-110.
[http://dx.doi.org/10.2337/dc20-S009]

[9] Edelman SV, Polonsky WH. Type 2 diabetes in the real world: the elusive nature of glycemic control. Diabetes Care 2017; 40(11): 1425-32.
[http://dx.doi.org/10.2337/dc16-1974] [PMID: 28801473]

[10] Kazemian P, Shebl FM, McCann N, Walensky RP, Wexler DJ. Evaluation of the cascade of diabetes care in the United States, 2005–2016. JAMA Intern Med 2019; 179(10): 1376-85.
[http://dx.doi.org/10.1001/jamainternmed.2019.2396] [PMID: 31403657]

[11] Khunti K, Ceriello A, Cos X, De Block C. Achievement of guideline targets for blood pressure, lipid, and glycaemic control in type 2 diabetes: A meta-analysis. Diabetes Res Clin Pract 2018; 137: 137-48.
[http://dx.doi.org/10.1016/j.diabres.2017.12.004] [PMID: 29325774]

[12] Chung WK, Erion K, Florez JC, *et al.* Precision medicine in diabetes: a consensus report from the American Diabetes Association (ADA) and the European Association for the Study of Diabetes (EASD). Diabetes Care 2020; 43(7): 1617-35.
[http://dx.doi.org/10.2337/dci20-0022] [PMID: 32561617]

[13] Inzucchi SE, Bergenstal RM, Buse JB, *et al.* Management of hyperglycemia in type 2 diabetes: a patient-centered approach: position statement of the American Diabetes Association (ADA) and the European Association for the Study of Diabetes (EASD). Diabetes Care 2012; 35(6): 1364-79.
[http://dx.doi.org/10.2337/dc12-0413] [PMID: 22517736]

[14] Gloyn AL, Drucker DJ. Precision medicine in the management of type 2 diabetes. Lancet Diabetes Endocrinol 2018; 6(11): 891-900.
[http://dx.doi.org/10.1016/S2213-8587(18)30052-4] [PMID: 29699867]

[15] Xie F, Chan JC, Ma RC. Precision medicine in diabetes prevention, classification and management. J Diabetes Investig 2018; 9(5): 998-1015.
[http://dx.doi.org/10.1111/jdi.12830] [PMID: 29499103]

[16] Finelli C, Sommella L, Gioia S, La Sala N, Tarantino G. Should visceral fat be reduced to increase longevity? Ageing Res Rev 2013; 12(4): 996-1004.
[http://dx.doi.org/10.1016/j.arr.2013.05.007] [PMID: 23764746]

[17] Sivakumar PM, Prabhawathi V, Zarrabi A, Akthar S, Prabhakar PK. Current Trends in the Therapeutic Strategies for Diabetes Management. Curr Med Chem 2021; 28(23): 4616-37.
[http://dx.doi.org/10.2174/0929867328666210218183914] [PMID: 33602069]

[18] Hauner H. The mode of action of thiazolidinediones. Diabetes Metab Res Rev 2002; 18 (Suppl. 2): S10-5.
[http://dx.doi.org/10.1002/dmrr.249] [PMID: 11921433]

[19] Greenfield JR, Chisholm DJ. Thiazolidinediones–mechanisms of action. Aust Prescr 2004; 27: 67-70.
[http://dx.doi.org/10.18773/austprescr.2004.059]

[20] Singh S, Loke YK. Thiazolidinediones and cardiovascular disease: balancing benefit and harm. Geriatr Aging 2008; 11: 179-83.

[21] Endo Y, Suzuki M, Yamada H, *et al.* Thiazolidinediones enhance sodium-coupled bicarbonate absorption from renal proximal tubules *via* PPARγ-dependent nongenomic signaling. Cell Metab 2011; 13(5): 550-61.

[http://dx.doi.org/10.1016/j.cmet.2011.02.015] [PMID: 21531337]

[22] Klip A, Leiter LA. Cellular mechanism of action of metformin. Diabetes Care 1990; 13(6): 696-704. [http://dx.doi.org/10.2337/diacare.13.6.696] [PMID: 2162756]

[23] Fowler MJ. Diabetes treatment, part 2: oral agents for glycemic management. Clin Diabetes 2007; 25: 131-4. [http://dx.doi.org/10.2337/diaclin.25.4.131]

[24] Proks P, Reimann F, Green N, Gribble F, Ashcroft F. Sulfonylurea stimulation of insulin secretion. Diabetes 2002; 51 (Suppl. 3): S368-76. [http://dx.doi.org/10.2337/diabetes.51.2007.S368] [PMID: 12475777]

[25] Aquilante CL. Sulfonylurea pharmacogenomics in Type 2 diabetes: the influence of drug target and diabetes risk polymorphisms. Expert Rev Cardiovasc Ther 2010; 8(3): 359-72. [http://dx.doi.org/10.1586/erc.09.154] [PMID: 20222815]

[26] Kalra S, Bahendeka S, Sahay R, *et al.* Consensus Recommendations on Sulfonylurea and Sulfonylurea Combinations in the Management of Type 2 Diabetes Mellitus - International Task Force. Indian J Endocrinol Metab 2018; 22(1): 132-57. [PubMed PMID: 29535952]. [http://dx.doi.org/10.4103/ijem.IJEM_556_17] [PMID: 29535952]

[27] Luna B, Feinglos MN. Oral agents in the management of type 2 diabetes mellitus. Am Fam Physician 2001; 63(9): 1747-56. [PMID: 11352285]

[28] Hekimsoy Z, Biberoğlu S, Cömlekçi A, Tarhan O, Mermut C, Biberoğlu K. Trimethoprim/sulfamethoxazole-induced hypoglycemia in a malnourished patient with severe infection. Eur J Endocrinol 1997; 136(3): 304-6. [http://dx.doi.org/10.1530/eje.0.1360304] [PMID: 9100556]

[29] Technology Appraisal Guidance NICE. 2013.https://www.nice.org.uk/guidance/ta288

[30] Inzucchi SE, Bergenstal RM, Buse JB, *et al.* Management of hyperglycemia in type 2 diabetes, 2015: a patient-centered approach: update to a position statement of the American Diabetes Association and the European Association for the Study of Diabetes. Diabetes Care 2015; 38(1): 140-9. [http://dx.doi.org/10.2337/dc14-2441] [PMID: 25538310]

[31] Handelsman Y, Bloomgarden ZT, Grunberger G, *et al.* American association of clinical endocrinologists and american college of endocrinology - clinical practice guidelines for developing a diabetes mellitus comprehensive care plan - 2015. Endocr Pract 2015; 21 (Suppl. 1): 1-87. [http://dx.doi.org/10.4158/EP15672.GLSUPPL] [PMID: 25869408]

[32] Nathan DM, Genuth S, Lachin J, *et al.* The effect of intensive treatment of diabetes on the development and progression of long-term complications in insulin-dependent diabetes mellitus. N Engl J Med 1993; 329(14): 977-86. [http://dx.doi.org/10.1056/NEJM199309303291401] [PMID: 8366922]

[33] Cherney DZI, Perkins BA, Soleymanlou N, *et al.* Renal hemodynamic effect of sodium-glucose cotransporter 2 inhibition in patients with type 1 diabetes mellitus. Circulation 2014; 129(5): 587-97. [http://dx.doi.org/10.1161/CIRCULATIONAHA.113.005081] [PMID: 24334175]

[34] Skrtić M, Yang GK, Perkins BA, *et al.* Characterisation of glomerular haemodynamic responses to SGLT2 inhibition in patients with type 1 diabetes and renal hyperfiltration. Diabetologia 2014; 57(12): 2599-602. [http://dx.doi.org/10.1007/s00125-014-3396-4] [PMID: 25280671]

[35] St Hilaire R, Costello H. Prescriber beware: report of adverse effect of sodium-glucose cotransporter 2 inhibitor use in a patient with contraindication. Am J Emerg Med 2014; 33: 604.e3-4. [http://dx.doi.org/10.1016/j.ajem.2014.09.039] [PMID: 25432591]

[36] Hayami T, Kato Y, Kamiya H, *et al.* Case of ketoacidosis by a sodium-glucose cotransporter 2 inhibitor in a diabetic patient with a low-carbohydrate diet. J Diabetes Investig 2015; 6(5): 587-90.

[http://dx.doi.org/10.1111/jdi.12330] [PMID: 26417418]

[37] Perkins BA, Cherney DZ, Partridge H, *et al.* Sodium-glucose cotransporter 2 inhibition and glycemic control in type 1 diabetes: results of an 8-week open-label proof-of-concept trial. Diabetes Care 2014; 37(5): 1480-3.
[http://dx.doi.org/10.2337/dc13-2338] [PMID: 24595630]

[38] FDA warns that SGLT2 inhibitors for diabetes may result in a serious condition of too much acid in the blood http://www.fda.gov/Drugs/DrugSafety/ucm446845.htm

[39] Review of diabetes medicines called SGLT2 inhibitors started http://www.ema.europa.eu/ema/index.jsp?curl=pages/medicines/human/referrals/SGLT2_inhibitors/human_referral_prac_000052.jsp&mid=WC0b01ac05805c516f

[40] Davis T, Edelman SV. Insulin therapy in type 2 diabetes. Med Clin North Am 2004; 88(4): 865-895, x.
[http://dx.doi.org/10.1016/j.mcna.2004.04.005] [PMID: 15308383]

[41] Miller CD, Phillips LS, Ziemer DC, Gallina DL, Cook CB, El-Kebbi IM. Hypoglycemia in patients with type 2 diabetes mellitus. Arch Intern Med 2001; 161(13): 1653-9.
[http://dx.doi.org/10.1001/archinte.161.13.1653] [PMID: 11434798]

[42] Giovannucci E, Harlan DM, Archer MC, *et al.* Diabetes and cancer: a consensus report. CA Cancer J Clin 2010; 60(4): 207-21.
[http://dx.doi.org/10.3322/caac.20078] [PMID: 20554718]

[43] Heslin MJ, Newman E, Wolf RF, Pisters PW, Brennan MF. Effect of systemic hyperinsulinemia in cancer patients. Cancer Res 1992; 52(14): 3845-50.
[PMID: 1617658]

[44] Giovannucci E. Metabolic syndrome, hyperinsulinemia, and colon cancer: a review. Am J Clin Nutr 2007; 86(3): s836-42.
[http://dx.doi.org/10.1093/ajcn/86.3.836S] [PMID: 18265477]

[45] Nauck MA, Meier JJ. The incretin effect in healthy individuals and those with type 2 diabetes: physiology, pathophysiology, and response to therapeutic interventions. Lancet Diabetes Endocrinol 2016; 4(6): 525-36.
[http://dx.doi.org/10.1016/S2213-8587(15)00482-9] [PMID: 26876794]

[46] Maruthur NM, Tseng E, Hutfless S, *et al.* Diabetes medications as monotherapy or metformin-based combination therapy for type 2 diabetes: a systematic review and meta-analysis. Ann Intern Med 2016; 164(11): 740-51.
[http://dx.doi.org/10.7326/M15-2650] [PMID: 27088241]

[47] American diabetes association. Physical activity/exercise and diabetes. Diabetes Care 2004; 27: 58-62.
[http://dx.doi.org/10.2337/diacare.27.2007.S58]

[48] Sigal RJ, Kenny GP, Wasserman DH, Castaneda-Sceppa C. Physical activity/exercise and type 2 diabetes. Diabetes Care 2004; 27(10): 2518-39.
[http://dx.doi.org/10.2337/diacare.27.10.2518] [PMID: 15451933]

[49] Umpierre D, Ribeiro PAB, Kramer CK, *et al.* Physical activity advice only or structured exercise training and association with HbA1c levels in type 2 diabetes: a systematic review and meta-analysis. JAMA 2011; 305(17): 1790-9.
[http://dx.doi.org/10.1001/jama.2011.576] [PMID: 21540423]

[50] Jeon CY, Lokken RP, Hu FB, van Dam RM, Van Dam RM. Physical activity of moderate intensity and risk of type 2 diabetes: a systematic review. Diabetes Care 2007; 30(3): 744-52.
[http://dx.doi.org/10.2337/dc06-1842] [PMID: 17327354]

[51] Boulé NG, Haddad E, Kenny GP, Wells GA, Sigal RJ. Effects of exercise on glycemic control and body mass in type 2 diabetes mellitus: a meta-analysis of controlled clinical trials. JAMA 2001; 286(10): 1218-27.
[http://dx.doi.org/10.1001/jama.286.10.1218] [PMID: 11559268]

[52] Ishii T, Yamakita T, Sato T, Tanaka S, Fujii S. Resistance training improves insulin sensitivity in NIDDM subjects without altering maximal oxygen uptake. Diabetes Care 1998; 21(8): 1353-5. [http://dx.doi.org/10.2337/diacare.21.8.1353] [PMID: 9702447]

[53] Cuff DJ, Meneilly GS, Martin A, Ignaszewski A, Tildesley HD, Frohlich JJ. Effective exercise modality to reduce insulin resistance in women with type 2 diabetes. Diabetes Care 2003; 26(11): 2977-82. [http://dx.doi.org/10.2337/diacare.26.11.2977] [PMID: 14578226]

[54] Dunstan DW, Daly RM, Owen N, *et al.* High-intensity resistance training improves glycemic control in older patients with type 2 diabetes. Diabetes Care 2002; 25(10): 1729-36. [http://dx.doi.org/10.2337/diacare.25.10.1729] [PMID: 12351469]

[55] Castaneda C, Layne JE, Munoz-Orians L, *et al.* A randomized controlled trial of resistance exercise training to improve glycemic control in older adults with type 2 diabetes. Diabetes Care 2002; 25(12): 2335-41. [http://dx.doi.org/10.2337/diacare.25.12.2335] [PMID: 12453982]

[56] Chang CL, Lin Y, Bartolome AP, Chen YC, Chiu SC, Yang WC. Herbal therapies for type 2 diabetes mellitus: chemistry, biology, and potential application of selected plants and compounds. Evid Based Complement Altern Med eCAM 2013; 378657. [http://dx.doi.org/10.1155/2013/378657]

[57] Mishra R. MohdShuaib S, Shiravan S, Mishra PS. A review on herbal antidiabetic drugs. J Appl Pharm Sci 2011; 01: 235-7.

[58] Kumar S, Anu M, Dinesh B, Amit M. Herbal medicines for diabetes management and its secondary complications. Curr Diabetes Rev 2020. [http://dx.doi.org/10.2174/1573399816666201103143225] [PMID: 33143632]

[59] Sarfraz M. Diabetes mellitus and Protective Approaches of Medicinal Plants: Present Status and Future Prospects (ISBN: 978-981-14-5911-5) Phytotherapy in the management of diabetes and hypertension. 2020; 3: pp. (23)38-60.

[60] Sarfraz M, Sajid S, Ashraf MA. Prevalence and pattern of dyslipidemia in hyperglycemic patients and its associated factors among Pakistani population. Saudi J Biol Sci 2016; 23(6): 761-6. [http://dx.doi.org/10.1016/j.sjbs.2016.03.001] [PMID: 27872574]

[61] Sarfraz M, Khaliq T, Khan JA, Aslam B. Effect of aqueous extract of black pepper and ajwa seed on liver enzymes in alloxan-induced diabetic Wister albino rats. Saudi Pharm J 2017; 25(4): 449-52. [http://dx.doi.org/10.1016/j.jsps.2017.04.004] [PMID: 28579873]

[62] Khaliq T, Sarfraz M, Ashraf MA. Recent Progress for the Utilization of *Curcuma longa, Piper nigrum* and *Phoenix dactylifera* Seeds against Type 2 Diabetes. West Indian Med J 2015; 64(5): 527-32. [PMID: 27399905]

[63] Patel DK, Prasad SK, Kumar R, Hemalatha S. An overview on antidiabetic medicinal plants having insulin mimetic property. Asian Pac J Trop Biomed 2012; 2(4): 320-30. [http://dx.doi.org/10.1016/S2221-1691(12)60032-X] [PMID: 23569923]

[64] Okoli CO, Obidike IC, Ezike AC, Akah PA, Salawu OA. Studies on the possible mechanisms of antidiabetic activity of extract of aerial parts of Phyllanthus niruri. Pharm Biol 2011; 49(3): 248-55. [http://dx.doi.org/10.3109/13880209.2010.501456] [PMID: 21323477]

[65] Hong SH, Heo JI, Kim JH, *et al.* Antidiabetic and Beta cell-protection activities of purple corn anthocyanins. Biomol Ther (Seoul) 2013; 21(4): 284-9. [http://dx.doi.org/10.4062/biomolther.2013.016] [PMID: 24244813]

[66] Allen RW, Schwartzman E, Baker WL, Coleman CI, Phung OJ. Cinnamon use in type 2 diabetes: an updated systematic review and meta-analysis. Ann Fam Med 2013; 11(5): 452-9. [http://dx.doi.org/10.1370/afm.1517] [PMID: 24019277]

[67] Baker WL, Gutierrez-Williams G, White CM, Kluger J, Coleman CI. Effect of cinnamon on glucose control and lipid parameters. Diabetes Care 2008; 31(1): 41-3. [http://dx.doi.org/10.2337/dc07-1711] [PMID: 17909085]

[68] Khan A, Safdar M, Ali Khan MM, Khattak KN, Anderson RA. Cinnamon improves glucose and lipids of people with type 2 diabetes. Diabetes Care 2003; 26(12): 3215-8. [http://dx.doi.org/10.2337/diacare.26.12.3215] [PMID: 14633804]

[69] Augusti KT, Sheela CG. Antiperoxide effect of S-allyl cysteine sulfoxide, an insulin secretagogue, in diabetic rats. Experientia 1996; 52(2): 115-20. [http://dx.doi.org/10.1007/BF01923354] [PMID: 8608811]

[70] Parastouei K, Ravanshad S, Mostaphavi H, Setoudehmaram E. Effects of garlic powder tablet on blood sugar, plasma lipids and blood pressure in type 2 diabetic patients with hyperlipidemia. Feyz J Kashan Uni Med Sci 2005; 9: 1-7.

[71] Ashraf R, Aamir K, Shaikh AR, Ahmed T. Effects of garlic on dyslipidemia in patients with type 2 diabetes mellitus. J Ayub Med Coll Abbottabad 2005; 17(3): 60-4. [PMID: 16320801]

[72] Sterti R. Berberine for diabetes mellitus type 2. Nat Med J 2010; 2: 5-6.

[73] Lee YS, Kim WS, Kim KH, *et al.* Berberine, a natural plant product, activates AMP-activated protein kinase with beneficial metabolic effects in diabetic and insulin-resistant states. Diabetes 2006; 55(8): 2256-64. [http://dx.doi.org/10.2337/db06-0006] [PMID: 16873688]

[74] Zhang Y, Li X, Zou D, *et al.* Treatment of type 2 diabetes and dyslipidemia with the natural plant alkaloid berberine. J Clin Endocrinol Metab 2008; 93(7): 2559-65. [http://dx.doi.org/10.1210/jc.2007-2404] [PMID: 18397984]

[75] Dong H, Wang N, Zhao L, Lu F. Berberine in the treatment of type 2 diabetes mellitus: a systemic review and meta-analysis. Evidence-Based Complementary Altern Med 2012; pp. 1-12.

[76] Meier JJ, Bhushan A, Butler PC. The potential for stem cell therapy in diabetes. Pediatr Res 2006; 59(4 Pt 2): 65R-73R. [http://dx.doi.org/10.1203/01.pdr.0000206857.38581.49] [PMID: 16549551]

[77] Butler AE, Janson J, Bonner-Weir S, Ritzel R, Rizza RA, Butler PC. Beta-cell deficit and increased beta-cell apoptosis in humans with type 2 diabetes. Diabetes 2003; 52(1): 102-10. [http://dx.doi.org/10.2337/diabetes.52.1.102] [PMID: 12502499]

[78] Pharmacologic approaches to glycemic treatment: Standards of Medical Care in Diabetes—2020. Diabetes Care 2020; 43 (Suppl. 1): S98-S110. [http://dx.doi.org/10.2337/dc20-S009] [PMID: 31862752]

[79] Atkinson MA, Bluestone JA, Eisenbarth GS, *et al.* How does type 1 diabetes develop?: the notion of homicide or β-cell suicide revisited. Diabetes 2011; 60(5): 1370-9. [http://dx.doi.org/10.2337/db10-1797] [PMID: 21525508]

[80] Hagopian W, Ferry RJ Jr, Sherry N, *et al.* Teplizumab preserves C-peptide in recent-onset type 1 diabetes: two-year results from the randomized, placebo-controlled Protégé trial. Diabetes 2013; 62(11): 3901-8. [http://dx.doi.org/10.2337/db13-0236] [PMID: 23801579]

[81] Gitelman SE, Gottlieb PA, Rigby MR, *et al.* Antithymocyte globulin treatment for patients with recent-onset type 1 diabetes: 12-month results of a randomised, placebo-controlled, phase 2 trial. Lancet Diabetes Endocrinol 2013; 1(4): 306-16. [http://dx.doi.org/10.1016/S2213-8587(13)70065-2] [PMID: 24622416]

[82] Mastrandrea L, Yu J, Behrens T, *et al.* Etanercept treatment in children with new-onset type 1 diabetes: pilot randomized, placebo-controlled, double-blind study. Diabetes Care 2009; 32(7): 1244-

9.
[http://dx.doi.org/10.2337/dc09-0054] [PMID: 19366957]

[83] Van Belle TL, Nierkens S, Arens R, von Herrath MG. Interleukin-21 receptor-mediated signals control autoreactive T cell infiltration in pancreatic islets. Immunity 2012; 36(6): 1060-72.
[http://dx.doi.org/10.1016/j.immuni.2012.04.005] [PMID: 22579473]

[84] Ryan GJ, Jobe LJ, Martin R. Pramlintide in the treatment of type 1 and type 2 diabetes mellitus. Clin Ther 2005; 27(10): 1500-12.
[http://dx.doi.org/10.1016/j.clinthera.2005.10.009] [PMID: 16330288]

[85] Ferrannini E. Sodium-glucose co-transporters and their inhibition: clinical physiology. Cell Metab 2017; 26(1): 27-38.
[http://dx.doi.org/10.1016/j.cmet.2017.04.011] [PMID: 28506519]

[86] Buse JB, Garg SK, Rosenstock J, *et al.* Sotagliflozin in combination with optimized insulin therapy in adults with type 1 diabetes: the North American inTandem1 study. Diabetes Care 2018; 41(9): 1970-80.
[http://dx.doi.org/10.2337/dc18-0343] [PMID: 29937430]

[87] Mathieu C, Zinman B, Hemmingsson JU, *et al.* Efficacy and safety of liraglutide added to insulin treatment in type 1 diabetes: the ADJUNCT ONE treat-to-target randomized trial. Diabetes Care 2016; 39(10): 1702-10.
[http://dx.doi.org/10.2337/dc16-0691] [PMID: 27506222]

[88] Xu G, Chen J, Jing G, Shalev A. Preventing β-cell loss and diabetes with calcium channel blockers. Diabetes 2012; 61(4): 848-56.
[http://dx.doi.org/10.2337/db11-0955] [PMID: 22442301]

[89] Ovalle F, Grimes T, Xu G, *et al.* Verapamil and beta cell function in adults with recent-onset type 1 diabetes. Nat Med 2018; 24(8): 1108-12.
[http://dx.doi.org/10.1038/s41591-018-0089-4] [PMID: 29988125]

[90] Abdi R, Fiorina P, Adra CN, Atkinson M, Sayegh MH. Immunomodulation by mesenchymal stem cells: a potential therapeutic strategy for type 1 diabetes. Diabetes 2008; 57(7): 1759-67.
[http://dx.doi.org/10.2337/db08-0180] [PMID: 18586907]

[91] Spaggiari GM, Abdelrazik H, Becchetti F, Moretta L. MSCs inhibit monocyte-derived DC maturation and function by selectively interfering with the generation of immature DCs: central role of MSC-derived prostaglandin E2. Blood 2009; 113(26): 6576-83.
[http://dx.doi.org/10.1182/blood-2009-02-203943] [PMID: 19398717]

[92] Wu J, Sun Z, Sun HS, *et al.* Intravenously administered bone marrow cells migrate to damaged brain tissue and improve neural function in ischemic rats. Cell Transplant 2008; 16(10): 993-1005.
[http://dx.doi.org/10.3727/000000007783472435] [PMID: 18351015]

[93] Tyndall A, Walker UA, Cope A, *et al.* Immunomodulatory properties of mesenchymal stem cells: a review based on an interdisciplinary meeting held at the Kennedy Institute of Rheumatology Division, London, UK, 31 October 2005. Arthritis Res Ther 2007; 9(1): 301.
[http://dx.doi.org/10.1186/ar2103] [PMID: 17284303]

[94] Voltarelli JC, Couri CEB, Stracieri ABPL, *et al.* Autologous nonmyeloablative hematopoietic stem cell transplantation in newly diagnosed type 1 diabetes mellitus. JAMA 2007; 297(14): 1568-76.
[http://dx.doi.org/10.1001/jama.297.14.1568] [PMID: 17426276]

[95] Maehr R, Chen S, Snitow M, *et al.* Generation of pluripotent stem cells from patients with type 1 diabetes. Proc Natl Acad Sci USA 2009; 106(37): 15768-73.
[http://dx.doi.org/10.1073/pnas.0906894106] [PMID: 19720998]

[96] Tang DQ, Cao LZ, Burkhardt BR, *et al.* *In vivo* and *in vitro* characterization of insulin-producing cells obtained from murine bone marrow. Diabetes 2004; 53(7): 1721-32.
[http://dx.doi.org/10.2337/diabetes.53.7.1721] [PMID: 15220196]

[97] Ianus A, Holz GG, Theise ND, Hussain MA. *In vivo* derivation of glucose-competent pancreatic endocrine cells from bone marrow without evidence of cell fusion. J Clin Invest 2003; 111(6): 843-50. [http://dx.doi.org/10.1172/JCI200316502] [PMID: 12639990]

[98] Ghannam S, Bouffi C, Djouad F, Jorgensen C, Noël D. Immunosuppression by mesenchymal stem cells: mechanisms and clinical applications. Stem Cell Res Ther 2010; 1(1): 2. [http://dx.doi.org/10.1186/scrt2] [PMID: 20504283]

[99] Chien KR, Moretti A, Laugwitz KL. Development. ES cells to the rescue. Science 2004; 306(5694): 239-40. [http://dx.doi.org/10.1126/science.1104769] [PMID: 15472069]

[100] Hori Y. Insulin-producing cells derived from stem/progenitor cells: therapeutic implications for diabetes mellitus. Med Mol Morphol 2009; 42(4): 195-200. [http://dx.doi.org/10.1007/s00795-009-0471-x] [PMID: 20033363]

[101] DiSanto RM, Subramanian V, Gu Z. Recent advances in nanotechnology for diabetes treatment. Wiley Interdiscip Rev Nanomed Nanobiotechnol 2015; 7(4): 548-64. [http://dx.doi.org/10.1002/wnan.1329] [PMID: 25641955]

[102] 102. Dhutia. Research paper based on pathology lectures at Medlink 2007 and Vet-Medlink 2007.

[103] Mo R, Jiang T, Di J, Tai W, Gu Z. Emerging micro- and nanotechnology based synthetic approaches for insulin delivery. Chem Soc Rev 2014; 43(10): 3595-629. [http://dx.doi.org/10.1039/c3cs60436e] [PMID: 24626293]

[104] Jiaojiao Y, Sun C, Wei Y, *et al.* Applying emerging technologies to improve diabetes treatment. Biomed Pharmacother 2018; 108: 1225-36. [http://dx.doi.org/10.1016/j.biopha.2018.09.155] [PMID: 30372824]

CHAPTER 3

Diabetes Type II: Should Aspartame be a Concern?

Arbind Kumar Choudhary[1]

[1] *Department of Physiology, All India Institute of Medical Science (AIIMS) Raebareli, Uttar Pradesh (U.P.), India*

Abstract: Blood sugar levels have to be controlled by individuals with type II diabetes (T2D) to preserve health and longevity. For such people, artificial sweeteners (including aspartame) are proposed sugar substitutes. In particular, the protection of aspartame has long been the point of discussion. Although it is such a problematic product, T2D patients are advised by many physicians to use it during a managed diet and as part of a treatment modality. Aspartame is 200 times sweeter than sugar and has a marginal effect on blood glucose levels. It is recommended for use so that T2D can regulate carbohydrate consumption and blood sugar levels. Previous studies, however, indicate that aspartame consumption may increase a person's risk of gaining weight instead of losing weight, resulting in intolerance to blood glucose in T2D. By increasing the levels of cortisol, aspartame can act as a biochemical stressor. It may cause systemic oxidative stress by creating excess free radicals, altering the gut's microbial activity, and interacting with the receptor N-methyl D-aspartate (NMDA), resulting in insulin deficiency or tolerance. Due to the lack of reliable evidence, aspartame and its derivatives are safe for T2D yet are still debatable. In the already stressful physiology of T2D, more research is needed to provide indications and raise concerns that aspartame may worsen the prevalence of pathological physiology.

Keyword: Aspartame, Aspartic acid, Methyl alcohol, Phenylalanine.

BACKGROUND

Non-nutritive sweeteners are commonly used by people who want to minimize their average daily calorie consumption, lose weight, and maintain a balanced diet [1]. Non-nutritive sweeteners elicited physiological responses, although inconsistent, but failed to reduce blood glucose levels [2]. Aspartame is a non-nutritive sweetener that has gotten a lot of attention because of its extreme sweetness, 200–300 times sweeter from sucrose [3]. The European Food Safety

* **Corresponding author Arbind Kumar Choudhary:** Department of Physiology, All India Institute of Medical Science (AIIMS) Raebareli, Uttar Pradesh (U.P.), India; E-mail: arbindchoudhary111@gmail.com

Shazia Anjum (Ed.)

Authority (EFSA) recommends 40 mg/kg.BW/day of aspartame, while the Food and Drug Administration (FDA) recommends 50 mg/kg.BW/day [4, 5]. Health-conscious people and diabetic patients use aspartame products, but their safety is a major concern (Fig. **1**).

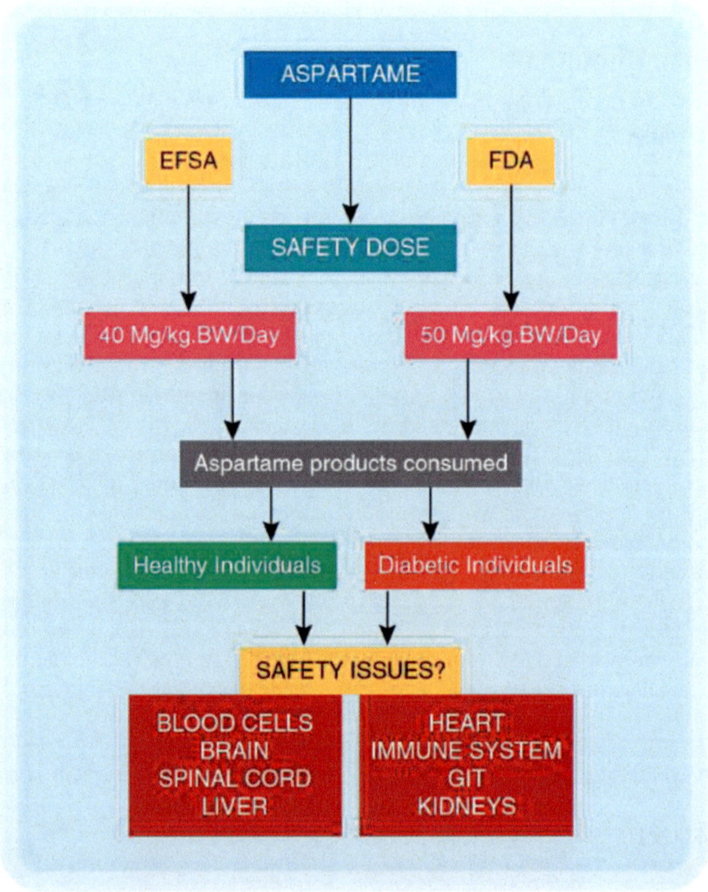

Fig. (1). Safety dosage of aspartame and safety issues.

Diabetes mellitus, *i.e.*, type II diabetes (T2D), is a metabolic disorder in which the pancreas fails to produce sufficient insulin. The body cells fail to respond to the insulin produced correctly. This results in chronic hyperglycemia (high blood glucose levels) and disturbances in carbohydrate, fat, and protein metabolism. In the long run, this may lead to symptoms of this disorder such as retinopathy, nephropathy; neuropathy; and an elevated risk of cardiovascular disease. A balanced diet, daily physical activity, and pharmacotherapy are all recommended for diabetes management. As for many people, the most critical component of the treatment regimen for diabetes is deciding on what to eat.

Aspartame and Weight Management

Although aspartame is suggested to help people lose weight by reducing their food intake and limiting their calorie intake [6], compared to natural sweeteners like sucrose, aspartame may have no impact on food consumption, satiety levels, or postprandial glucose levels. It may also not affect postprandial insulin levels [7]. Although aspartame can help with weight loss by lowering caloric intake when compared to sucrose [6], there is evidence that rats can compensate for the calorie reduction by overeating, resulting in increased body weight and adiposity [8]. It is well known that type 2 diabetes (T2D) and obesity have a troubling relationship [9].

Humans with a higher BMI were found to consume diet carbonated drinks containing aspartame [10, 11]. The increasing use of aspartame (*e.g.*, Diet Coke) in food items has been related to weight gain [12]. Aspartame is thought to disrupt appetite control and contribute to weight gain. It does not stimulate the food reward pathways in the same way that natural sweeteners do but instead encourages sugar craving and sugar dependency, leading to weight gain [12]. For some people, eating dietary foods justifies consuming excess calories from other kinds of food. As a result, it's impossible to say if obesity is linked to the usage of artificial sweeteners (including aspartame) or just to eat too many calories [13].

Weight changes are typically connected to insulin receptors or insulin resistance changes [14]. Increased insulin and glucose levels are linked to weight gain [15]. Chronically high insulin levels are linked to a loss of insulin sensitivity [16], leading to insulin resistance [14]. Insulin resistance is believed to be related to elevated blood sugar, triglycerides, blood clots, insomnia, and cardiovascular and neurological disease [17 - 19].

Although replacing added sugars in foods and beverages with aspartame has the potential to improve body weight and glucose control. The American Diabetes Association and the American Heart Association said in a scientific statement that evidence for their long-term benefits in reducing caloric and added sugar intake is limited.

Aspartame and Glucose Intolerance

Glucose intolerance is commonly accepted as a precursor to T2D [20]. In human [7, 21, 22] and animal studies [23-25], the role of aspartame in maintaining an average blood glucose level is debatable. Although no major variations in blood glucose levels were found [7, 22], it did not sustain an average level. It increased blood glucose levels [23, 25, 26]. Gut enzymes (esterase and peptidase) easily break down aspartame into its three metabolic components: phenylalanine (50%), aspartate (40%), and methanol (10%) [27]. Aspartame and its metabolites have been linked to blood glucose dysregulation [28]: (1) Neuroendocrine imbalances are disrupted; (2) the N-methyl D-aspartate (NMDA) receptor is altered; (3) liver function is impaired, and (4) gut microbes are altered.

Glucose homeostasis is maintained by the neuro-endocrine system [29]. The liver, pancreas, and brain all have glucose receptors (GLUTS) [30]. The hypothalamic-pituitary-adrenal (HPA) axis regulates glucose homeostasis [29]. Aspartame is a chemical stressor that causes excess corticosterone (cortisol) production in the hypothalamic-pituitary-adrenal (HPA) axis [31]. Hyperglycemia and insulin resistance can result from a disturbance in glucose homeostasis [32]. Aspartate, a component of aspartame, has been shown to activate the NMDA receptor and occupy glutamate binding sites [27]. Central excitatory amino acids activate NMDA receptors in the hypothalamic-pituitary-adrenal axis during hypoglycemia, resulting in stimulation of the sympathoadrenal and hypothalamic–pituitary–adrenal axis and appear to play an essential role in the sustained elevation in hepatic glucose output [33] and as a consequence, consuming aspartame-sweetened beverages while hypoglycemic can interfere with the glucoregulatory response. During fasting and after feeding, the liver retains normal glucose levels, and it is an essential site of insulin clearance [34]. When insulin is absent or the liver is insulin resistant, hepatic glucose production and glycogenolysis can cause hyperglycemia [35]. At the recommended dose of 40 mg/kg.BW/day, aspartame can cause abnormal hepatocellular function [25, 36 - 39]. Hepatic dysfunction is linked to decreased hepatic insulin sensitivity and a reduction in blood glucose levels [40].

Gut microbes influence the primary host biological systems that regulate energy homeostasis and glucose metabolism in T2D [41] and play a key role in insulin resistance production [42]. In diet-induced obese rats, low dose aspartame (5–7 mg/kg/day) in drinking water caused elevated fasting glucose levels and impaired insulin tolerance, and fecal analysis of gut microbiota revealed aspartame increased the abundance of Enterobacteriaceae [43]. Mice that drank water containing 4% aspartame and ate a high-fat diet had higher glucose excursions after a glucose load was linked to a metabolic phenotype shift triggered by changes in the gut microbiota [26]. A dysregulated microbiota-gut-brain axis may explain aspartame metabolic and other side-effects [44].

“At this time, there is insufficient data to determine conclusively whether the use of aspartame to replace caloric sweeteners in beverages and foods reduces added sugars or carbohydrate intakes, or benefits appetite, energy balance, body weight, or cardiometabolic risk factors.”

Aspartame and Insulin Resistance

Insulin resistance is a state of impaired biological response to normal or elevated serum insulin concentrations [45] and occurs when the body does not respond appropriately to insulin. Insulin resistance may also be the cause of abnormally high blood glucose levels in T2D [46] due to (a) reduced early insulin secretory response to oral glucose, (b) decreased glucose-sensing ability of the cell, (c) reduced the ability of the cell to compensate for the degree of insulin resistance.

In general, cortisol has been linked to insulin resistance through the following mechanism: (1). Cortisol decreases the translocation of GLUT-4 transports and associated glucose uptake [47, 48]. (2). Cortisol inhibits insulin release from the beta cells of the pancreas in mice [49]. (3). Cortisol facilitates insulin resistance by increasing glucose production and accumulation of lipids in the cell [47, 48]. Hence excess cortisol after chronic aspartame consumption may promote insulin resistance. However, the particular mechanism has to be explored further with more scientific studies.

Aspartame (0.625-45mg/kg) consumption may exert a dose-dependent inhibition of brain serotonin, noradrenaline, and dopamine [50] in a changed neurological function. Phenylalanine, an aspartame component, competes with tryptophan, the serotonin precursor, for the same channel (NAAT) through the blood-brain barrier [27]. Phenylalanine penetrates the brain and suppresses serotonin levels [27]. People with low levels of serotonin are often compelled to consume more sugar in a bid to increase serotonin production, and this often results in a sugar addiction [51], which in turn can lead to insulin resistance (high levels of insulin cause receptors for insulin to shut down through 'down-regulation) [52].

Systemic oxidative stress is associated with insulin resistance [53]. Aspartame induces excess free radical production, particularly reactive oxygen species (ROS) and reactive nitrogen species (RNS). These free radicals result in systemic oxidative stress [54]. ROS/RNS may impair insulin signaling [55 - 57] by (a) inducing IRS serine/threonine phosphorylation, (b) upsetting cellular redistribution of insulin signaling components, (c) declining GLUT4 gene transcription or (d) altering mitochondrial activity. Increased oxidative stress may cause insulin resistance by inhibiting insulin signals and deregulating adiponectin [58, 59] and other adipocyte-derived factors such as TNF-α [60], leptin [61], and free fatty acids (FFAs) [62]. Hence, systemic oxidative stress induced by aspartame usage may exacerbate insulin resistance and impaired glucose tolerance, and increase complications in T2D.

The European Food Safety Authority (EFSA) [63] Panel used a Weight-of-Data approach for aspartame and combined with an examination of the biological significance of the appraised and validated evidence, followed by an uncertainty analysis. Finally, an examination of the distribution of negative *versus* positive study outcomes in terms of dependability revealed that the assertion of bias in aspartame scientific risk assessment is unfounded.

The Academy of Nutrition and Dietetics, the American Heart Association, and the American Diabetes Association state that aspartame should be used with caution when consumed as part of a healthy diet with following current nutrition guidelines [64].

In general, experts advise using aspartame in the least amount is feasible. "On a single-dose basis, these sweeteners are safe.

CONCLUSION

Aspartame used in T2D may lead to weight gain rather than weight loss. Aspartame consumption may act as a chemical stressor, increasing cortisol levels, which interfere with insulin pathways. Moreover, aspartame consumption may induce systemic oxidative stress by producing excess free radicals, leading to inflammation that may exacerbate sugar control in T2D complications. Due to the large diversity of study types and designs, no strong conclusion about whether aspartame intake is directly connected with T2D or whether it has beneficial or negative effects on the illness. There is an urgent need for well-designed studies that encompass all types of potentially interfering factors as well as those that play a role in T2D development.

We urge all national and international public health agencies to re-evaluate their estimates of aspartame's health concerns. It is the responsibility of health-care practitioners to keep up to date on the current evidence-based dietary guidelines and to warn customers about the potential hazards of using aspartame.

CONSENT FOR PUBLICATION

Not applicable.

CONFLICT OF INTEREST

The author declares no conflict of interest, financial or otherwise.

ACKNOWLEDGEMENTS

Declared none.

REFERENCES

[1] Burke MV, Small DM. Physiological mechanisms by which non-nutritive sweeteners may impact body weight and metabolism. Physiol Behav 2015; 152(Pt B): 381-8. [http://dx.doi.org/10.1016/j.physbeh.2015.05.036] [PMID: 26048305]

[2] Tucker RM, Tan S-Y. Do non-nutritive sweeteners influence acute glucose homeostasis in humans? A systematic review. Physiol Behav 2017; 182: 17-26. [http://dx.doi.org/10.1016/j.physbeh.2017.09.016] [PMID: 28939430]

[3] Marinovich M, Galli CL, Bosetti C, Gallus S, La Vecchia C. Aspartame, low-calorie sweeteners and disease: regulatory safety and epidemiological issues. Food Chem Toxicol 2013; 60: 109-15. [http://dx.doi.org/10.1016/j.fct.2013.07.040] [PMID: 23891579]

[4] Duffy VB, Anderson GH. Position of the American Dietetic Association: use of nutritive and nonnutritive sweeteners. J Am Diet Assoc 1998; 98(5): 580-7. [http://dx.doi.org/10.1016/S0002-8223(98)00131-X] [PMID: 9597035]

[5] Butchko HH, Kotsonis FN. Acceptable daily intake *vs* actual intake: the aspartame example. J Am Coll Nutr 1991; 10(3): 258-66. [http://dx.doi.org/10.1080/07315724.1991.10718153] [PMID: 1894884]

[6] De la Hunty A, Gibson S, Ashwell M. A review of the effectiveness of aspartame in helping with weight control. Nutr Bull 2006; 31(2): 115-28. [http://dx.doi.org/10.1111/j.1467-3010.2006.00564.x]

[7] Anton SD, Martin CK, Han H, *et al.* Effects of stevia, aspartame, and sucrose on food intake, satiety, and postprandial glucose and insulin levels. Appetite 2010; 55(1): 37-43. [http://dx.doi.org/10.1016/j.appet.2010.03.009] [PMID: 20303371]

[8] Feijó FM, Ballard CR, Foletto KC, *et al.* Saccharin and aspartame, compared with sucrose, induce greater weight gain in adult Wistar rats, at similar total caloric intake levels. Appetite 2013; 60(1): 203-7. [http://dx.doi.org/10.1016/j.appet.2012.10.009] [PMID: 23088901]

[9] Swinburn BA, Caterson I, Seidell JC, James WP. Diet, nutrition and the prevention of excess weight gain and obesity. Public Health Nutr 2004; 7(1A): 123-46. [http://dx.doi.org/10.1079/PHN2003585] [PMID: 14972057]

[10] Blum JW, Jacobsen DJ, Donnelly JE. Beverage consumption patterns in elementary school aged children across a two-year period. J Am Coll Nutr 2005; 24(2): 93-8. [http://dx.doi.org/10.1080/07315724.2005.10719449] [PMID: 15798075]

[11] Forshee RA, Storey ML. Total beverage consumption and beverage choices among children and adolescents. Int J Food Sci Nutr 2003; 54(4): 297-307. [http://dx.doi.org/10.1080/09637480120092143] [PMID: 12850891]

[12] Yang Q. Gain weight by "going diet?" Artificial sweeteners and the neurobiology of sugar cravings: Neuroscience 2010. Yale J Biol Med 2010; 83(2): 101-8. [PMID: 20589192]

[13] Kvaavik E, Andersen LF, Klepp K-I. The stability of soft drinks intake from adolescence to adult age and the association between long-term consumption of soft drinks and lifestyle factors and body weight. Public Health Nutr 2005; 8(2): 149-57. [http://dx.doi.org/10.1079/PHN2004669] [PMID: 15877908]

[14] Elliott SS, Keim NL, Stern JS, Teff K, Havel PJ. Fructose, weight gain, and the insulin resistance syndrome. Am J Clin Nutr 2002; 76(5): 911-22. [http://dx.doi.org/10.1093/ajcn/76.5.911] [PMID: 12399260]

[15] Lazzer S, Vermorel M, Montaurier C, Meyer M, Boirie Y. Changes in adipocyte hormones and lipid oxidation associated with weight loss and regain in severely obese adolescents. Int J Obes 2005; 29(10): 1184-91.

[http://dx.doi.org/10.1038/sj.ijo.0802977] [PMID: 16155586]

[16] Reinehr T, Kiess W, Kapellen T, Andler W. Insulin sensitivity among obese children and adolescents, according to degree of weight loss. Pediatrics 2004; 114(6): 1569-73. [http://dx.doi.org/10.1542/peds.2003-0649-F] [PMID: 15574616]

[17] Basoglu OK, Sarac F, Sarac S, Uluer H, Yilmaz C. Metabolic syndrome, insulin resistance, fibrinogen, homocysteine, leptin, and C-reactive protein in obese patients with obstructive sleep apnea syndrome. Ann Thorac Med 2011; 6(3): 120-5. [http://dx.doi.org/10.4103/1817-1737.82440] [PMID: 21760842]

[18] Bonora E, Kiechl S, Willeit J, *et al.* Insulin resistance as estimated by homeostasis model assessment predicts incident symptomatic cardiovascular disease in caucasian subjects from the general population: the Bruneck study. Diabetes Care 2007; 30(2): 318-24. [http://dx.doi.org/10.2337/dc06-0919] [PMID: 17259501]

[19] Kim B, Feldman EL. Insulin resistance in the nervous system. Trends Endocrinol Metab 2012; 23(3): 133-41. [http://dx.doi.org/10.1016/j.tem.2011.12.004] [PMID: 22245457]

[20] Bastard J-P, Maachi M, Lagathu C, *et al.* Recent advances in the relationship between obesity, inflammation, and insulin resistance. Eur Cytokine Netw 2006; 17(1): 4-12. [PMID: 16613757]

[21] Melanson KJ, Westerterp-Plantenga MS, Campfield LA, Saris WH. Blood glucose and meal patterns in time-blinded males, after aspartame, carbohydrate, and fat consumption, in relation to sweetness perception. Br J Nutr 1999; 82(6): 437-46. [http://dx.doi.org/10.1017/S0007114599001695] [PMID: 10690159]

[22] Wilson F, Howes K. Blood glucose changes following the ingestion of sucrose-and aspartame-sweetened beverages. Appetite 2008; 51(2): 410. [http://dx.doi.org/10.1016/j.appet.2008.04.261] [PMID: 19110019]

[23] Collison KS, Makhoul NJ, Zaidi MZ, *et al.* Interactive effects of neonatal exposure to monosodium glutamate and aspartame on glucose homeostasis. Nutr Metab (Lond) 2012; 9(1): 58. [http://dx.doi.org/10.1186/1743-7075-9-58] [PMID: 22697049]

[24] Collison KS, Makhoul NJ, Zaidi MZ, *et al.* Gender dimorphism in aspartame-induced impairment of spatial cognition and insulin sensitivity. PLoS One 2012; 7(4): e31570. [http://dx.doi.org/10.1371/journal.pone.0031570] [PMID: 22509243]

[25] Kim J-Y, Seo J, Cho K-H. Aspartame-fed zebrafish exhibit acute deaths with swimming defects and saccharin-fed zebrafish have elevation of cholesteryl ester transfer protein activity in hypercholesterolemia. Food Chem Toxicol 2011; 49(11): 2899-905. [http://dx.doi.org/10.1016/j.fct.2011.08.001] [PMID: 21855599]

[26] Suez J, Korem T, Zeevi D, *et al.* Artificial sweeteners induce glucose intolerance by altering the gut microbiota. Nature 2014; 514(7521): 181-6. [http://dx.doi.org/10.1038/nature13793] [PMID: 25231862]

[27] Humphries P, Pretorius E, Naudé H. Direct and indirect cellular effects of aspartame on the brain. Eur J Clin Nutr 2008; 62(4): 451-62. [http://dx.doi.org/10.1038/sj.ejcn.1602866] [PMID: 17684524]

[28] Choudhary AK. Aspartame: should individuals with Type II Diabetes be taking it? Curr Diabetes Rev 2018; 14(4): 350-62. [http://dx.doi.org/10.2174/1573399813666170601093336] [PMID: 28571543]

[29] Nadal A, Alonso-Magdalena P, Soriano S, Quesada I, Ropero AB. The pancreatic β-cell as a target of estrogens and xenoestrogens: Implications for blood glucose homeostasis and diabetes. Mol Cell Endocrinol 2009; 304(1-2): 63-8. [http://dx.doi.org/10.1016/j.mce.2009.02.016] [PMID: 19433249]

[30] Nishiumi S, Bessyo H, Kubo M, *et al.* Green and black tea suppress hyperglycemia and insulin resistance by retaining the expression of glucose transporter 4 in muscle of high-fat diet-fed C57BL/6J mice. J Agric Food Chem 2010; 58(24): 12916-23.
[http://dx.doi.org/10.1021/jf102840w] [PMID: 21105694]

[31] Iyyaswamy A, Rathinasamy S. Effect of chronic exposure to aspartame on oxidative stress in the brain of albino rats. J Biosci 2012; 37(4): 679-88.
[http://dx.doi.org/10.1007/s12038-012-9236-0] [PMID: 22922192]

[32] Jung HS, Chung KW, Won Kim J, *et al.* Loss of autophagy diminishes pancreatic β cell mass and function with resultant hyperglycemia. Cell Metab 2008; 8(4): 318-24.
[http://dx.doi.org/10.1016/j.cmet.2008.08.013] [PMID: 18840362]

[33] Molina PE, Abumrad NN. Contribution of excitatory amino acids to hypoglycemic counter-regulation. Brain Res 2001; 899(1-2): 201-8.
[http://dx.doi.org/10.1016/S0006-8993(01)02236-3] [PMID: 11311881]

[34] Kalsbeek A, la Fleur S, Fliers E. Circadian control of glucose metabolism. Mol Metab 2014; 3(4): 372-83.
[http://dx.doi.org/10.1016/j.molmet.2014.03.002] [PMID: 24944897]

[35] Rojas JM, Schwartz MW. Control of hepatic glucose metabolism by islet and brain. Diabetes Obes Metab 2014; 16(S1) (Suppl. 1): 33-40.
[http://dx.doi.org/10.1111/dom.12332] [PMID: 25200294]

[36] Ashok I, Sheeladevi R. Oxidant stress evoked damage in rat hepatocyte leading to triggered nitric oxide synthase (NOS) levels on long term consumption of aspartame. J Food Drug Anal 2015; 23(4): 679-91.
[http://dx.doi.org/10.1016/j.jfda.2014.07.011] [PMID: 28911484]

[37] Choudhary AK, Devi RS. Serum biochemical responses under oxidative stress of aspartame in wistar albino rats. Asian Pac J Trop Dis 2014; 4: S403-10.
[http://dx.doi.org/10.1016/S2222-1808(14)60478-3]

[38] Iman MM. Effect of aspartame on some oxidative stress parameters in liver and kidney of rats. Afr J Pharm Pharmacol 2011; 5(6): 678-82.
[http://dx.doi.org/10.5897/AJPP10.302]

[39] Kumar Choudhary A, Selvaraj S, Sheela Devi R. Aspartame Induce Modification in Membrane Bound and Antioxidant Enzymes in Liver and Kidney of Wistar Albino Rats. Curr Nutr Food Sci 2014; 10(4): 275-87.
[http://dx.doi.org/10.2174/1573401310666141024215431]

[40] Nguyen MT, Favelyukis S, Nguyen A-K, *et al.* A subpopulation of macrophages infiltrates hypertrophic adipose tissue and is activated by free fatty acids *via* Toll-like receptors 2 and 4 and JNK-dependent pathways. J Biol Chem 2007; 282(48): 35279-92.
[http://dx.doi.org/10.1074/jbc.M706762200] [PMID: 17916553]

[41] Cani PD, Geurts L, Matamoros S, Plovier H, Duparc T. Glucose metabolism: focus on gut microbiota, the endocannabinoid system and beyond. Diabetes Metab 2014; 40(4): 246-57.
[http://dx.doi.org/10.1016/j.diabet.2014.02.004] [PMID: 24631413]

[42] Shen J, Obin MS, Zhao L. The gut microbiota, obesity and insulin resistance. Mol Aspects Med 2013; 34(1): 39-58.
[http://dx.doi.org/10.1016/j.mam.2012.11.001] [PMID: 23159341]

[43] Palmnäs MS, Cowan TE, Bomhof MR, *et al.* Low-dose aspartame consumption differentially affects gut microbiota-host metabolic interactions in the diet-induced obese rat. PLoS One 2014; 9(10): e109841.
[http://dx.doi.org/10.1371/journal.pone.0109841] [PMID: 25313461]

[44] Choudhary AK, Lee Y. Dysregulated microbiota-gut-brain axis: Does it explain aspartame metabolic

and other side-effects? Nutr Food Sci 2017; 47(5): 648-58.
[http://dx.doi.org/10.1108/NFS-03-2017-0034]

[45] Wilcox G. Insulin and insulin resistance. Clin Biochem Rev 2005; 26(2): 19-39.
[PMID: 16278749]

[46] Polonsky KS, Sturis J, Bell GI. Seminars in Medicine of the Beth Israel Hospital, Boston. Non-insuli--dependent diabetes mellitus - a genetically programmed failure of the beta cell to compensate for insulin resistance. N Engl J Med 1996; 334(12): 777-83.
[http://dx.doi.org/10.1056/NEJM199603213341207] [PMID: 8592553]

[47] Andrews RC, Walker BR. Glucocorticoids and insulin resistance: old hormones, new targets. Clin Sci (Lond) 1999; 96(5): 513-23.
[http://dx.doi.org/10.1042/CS19980388] [PMID: 10209084]

[48] Reynolds RM, Walker BR. Human insulin resistance: the role of glucocorticoids. Diabetes Obes Metab 2003; 5(1): 5-12.
[http://dx.doi.org/10.1046/j.1463-1326.2003.00221.x] [PMID: 12542720]

[49] Delaunay F, Khan A, Cintra A, *et al.* Pancreatic beta cells are important targets for the diabetogenic effects of glucocorticoids. J Clin Invest 1997; 100(8): 2094-8.
[http://dx.doi.org/10.1172/JCI119743] [PMID: 9329975]

[50] Abdel-Salam OM, Salem NA, Hussein JS. Effect of aspartame on oxidative stress and monoamine neurotransmitter levels in lipopolysaccharide-treated mice. Neurotox Res 2012; 21(3): 245-55.
[http://dx.doi.org/10.1007/s12640-011-9264-9] [PMID: 21822758]

[51] Volkow ND, Wang G-J, Baler RD. Reward, dopamine and the control of food intake: implications for obesity. Trends Cogn Sci 2011; 15(1): 37-46.
[http://dx.doi.org/10.1016/j.tics.2010.11.001] [PMID: 21109477]

[52] Luo S, Luo J, Cincotta AH. Chronic ventromedial hypothalamic infusion of norepinephrine and serotonin promotes insulin resistance and glucose intolerance. Neuroendocrinology 1999; 70(6): 460-5.
[http://dx.doi.org/10.1159/000054508] [PMID: 10657739]

[53] Meigs JB, Larson MG, Fox CS, Keaney JF Jr, Vasan RS, Benjamin EJ. Association of oxidative stress, insulin resistance, and diabetes risk phenotypes: the Framingham Offspring Study. Diabetes Care 2007; 30(10): 2529-35.
[http://dx.doi.org/10.2337/dc07-0817] [PMID: 17586736]

[54] Choudhary AK, Pretorius E. Revisiting the safety of aspartame. Nutr Rev 2017; 75(9): 718-30.
[http://dx.doi.org/10.1093/nutrit/nux035] [PMID: 28938797]

[55] Bloch-Damti A, Bashan N. Proposed mechanisms for the induction of insulin resistance by oxidative stress. Antioxid Redox Signal 2005; 7(11-12): 1553-67.
[http://dx.doi.org/10.1089/ars.2005.7.1553] [PMID: 16356119]

[56] Morino K, Petersen KF, Shulman GI. Molecular mechanisms of insulin resistance in humans and their potential links with mitochondrial dysfunction. Diabetes 2006; 55(2) (Suppl. 2): S9-S15.
[http://dx.doi.org/10.2337/db06-S002] [PMID: 17130651]

[57] Rains JL, Jain SK. Oxidative stress, insulin signaling, and diabetes. Free Radic Biol Med 2011; 50(5): 567-75.
[http://dx.doi.org/10.1016/j.freeradbiomed.2010.12.006] [PMID: 21163346]

[58] Furukawa S, Fujita T, Shimabukuro M, *et al.* Increased oxidative stress in obesity and its impact on metabolic syndrome. J Clin Invest 2004; 114(12): 1752-61.
[http://dx.doi.org/10.1172/JCI21625] [PMID: 15599400]

[59] Houstis N, Rosen ED, Lander ES. Reactive oxygen species have a causal role in multiple forms of insulin resistance. Nature 2006; 440(7086): 944-8.
[http://dx.doi.org/10.1038/nature04634] [PMID: 16612386]

[60] Hotamisligil GS, Spiegelman BM. Tumor necrosis factor α: a key component of the obesity-diabetes link. Diabetes 1994; 43(11): 1271-8.
[http://dx.doi.org/10.2337/diab.43.11.1271] [PMID: 7926300]

[61] Cohen B, Novick D, Rubinstein M. Modulation of insulin activities by leptin. Science 1996; 274(5290): 1185-8.
[http://dx.doi.org/10.1126/science.274.5290.1185] [PMID: 8895466]

[62] McGarry JD. Banting lecture 2001: dysregulation of fatty acid metabolism in the etiology of type 2 diabetes. Diabetes 2002; 51(1): 7-18.
[http://dx.doi.org/10.2337/diabetes.51.1.7] [PMID: 11756317]

[63] Kass GEN, Lodi F. Letter to the editor regarding the article 'EFSA's toxicological assessment of aspartame: was it even-handedly trying to identify possible unreliable positives and unreliable negatives?'. Arch Public Health 2020; 78(1): 14.
[http://dx.doi.org/10.1186/s13690-020-0395-4] [PMID: 32266067]

[64] Landrigan PJ, Straif K. Aspartame and cancer - new evidence for causation. Environ Health 2021; 20(1): 42.
[http://dx.doi.org/10.1186/s12940-021-00725-y] [PMID: 33845854]

Frontiers in Clinical Drug Research-Diabetes & Obesity, 2023, *Vol. 7*, 59-116

CHAPTER 4

Mental Health, Adherence, and Self-Management Among Children with Diabetes

Beáta Erika Nagy[1], **Brigitta Munkácsi**[1] and **Karolina Eszter Kovács**[2,*]

[1] *University of Debrecen, Faculty of Medicine, Institute of Pediatrics, Pediatric Psychology and Psychosomatic Unit, Hungary*

[2] *Faculty of Arts, Institute of Psychology, Department of Pedagogical Psychology, Hungary*

Abstract: Nowadays, the investigation of mental health is a popular and important topic. Several national and international researchers have been trying to discover the different mechanisms, effects and efficacy among healthy people and patients diagnosed with chronic diseases. It is particularly important to monitor this phenomenon in childhood and adolescence regularly. The developmental processes are further hampered by the physical, mental, social and spiritual development due to the different illnesses. Therefore, it is clear that mapping mental health and various therapeutic procedures, as well as their positive and negative effects, are of paramount importance in diabetes and obesity.

In this research, after analysing the scales of ten international questionnaires, a complex Diabetes Adherence Questionnaire with 58 statements was created, the characteristics and subscales of which (1. Self-management; 2. Emotional feedback - emotional reactions associated with blood sugar level measurement; 3. Social support - parents and family; 4. Social support - peer relationships; 5. Denial of the disease; 6. Positive consequences of adherence; 7. Negative consequences of adherence, pain, discomfort, burden; 8. Relationship with the medical team; 9. Concern about the future) are described in the present book chapter. We also introduce our latest research findings on the relationship between adherence and mental health, covering self-evaluated health and quality of life, satisfaction with life, subjective well-being, vision and depression, stating that positive variables show a positive while negative variables correlate negatively with adherence.

Keywords: T1DM, Adherence, Denial of the disease, Depression,, Diabetes Adherence Questionnaire, Emotional feedback, Negative adherence (the burden of the treatment), Positive adherence, Quality of life, Self-management, Self-rated

* **Corresponding author Karolina Eszter Kovács:** Faculty of Arts, Institute of Psychology, Department of Pedagogical Psychology, Hungary; Tel: +36 52 512 900/22533; E-mail: karolina92.kovacs@gmail.com

Shazia Anjum (Ed.)

health, Social support (medical team) vision (worries), Social support (parents and family), Social support (peer relationships).

INTRODUCTION

According to the latest statistics of the International Diabetes Atlas, Type 1 Diabetes (T1DM) is one of the fastest-growing global health problems of the 21st century [1, 2]. Epidemiological surveys show that its incidence and prevalence are continuously increasing worldwide, affecting all age groups, regardless of gender and socio-economic background. As diabetes has a significant impact on children's physical health and their mental, emotional, and social development [3, 4], continuous and in-depth exploration of T1DM and related factors is of paramount importance. Chronic diseases such as diabetes require adequate adherence to the treatment protocol, in this case, regular insulin dosage, blood glucose measurement, and proper diet [5]. However, its quality can be supported or hindered by several intra- and interpersonal as well as environmental factors [6]. Adherence, which is 'the individual's behaviour in accordance with recommendations agreed with a health care professional in medication, diet, and lifestyle change', is thus a complex phenomenon that also requires a complex definition to study. However, the questionnaires and other research methods applied in international practice to study adherence do not cover adherence in complexity but focus only on one spectrum. Thus, we aimed to create a complex Diabetes Adherence Questionnaire with 58 statements [7]. In this chapter, after introducing the most relevant literature and previous research findings, we present the above-mentioned questionnaire and the most important findings of these topics.

MENTAL HEALTH AND T1DM

Quality Of Health And Diabetes

The concept of quality of life (QoL) has come to the fore in psychology and medicine in recent decades. The term quality of life, interpreted from a psychological point of view, is based on positive psychology and is associated with the subjective well-being and the affective dimension of quality of life [8, 9]. In addition to the general satisfaction, the cognitive components of the quality of life also mean an area-specific assessment related to individual satisfaction, performance, and health [10]. Quality of life is determined by the subjective assessment of the individuals' life and how good or bad they feel about it. Thus, the multidimensional construct that integrates physical, psychological, and social well-being includes both cognitive and emotional elements [11, 12]. First, the study and improvement of quality of life among children with certain somatic diseases, *e.g.* diabetes, cardiac disease and epilepsy, have appeared. Concern-ing the quality of life, the subjective assessment of an individual's general health, impairments, and routine functioning are significant [13]. When examining the phenomenon, the aspect described for adults is of outstanding importance, according to which an objective external observer is essential in addition to the child's own judgment, so we cannot rely only on the children's subjective evaluation. The use of proxy reports, *i.e.* data based on the opinion of the external reviewer (mostly the parent), is recommended to get a more precise and reliable picture of the situation of children and adolescents. However, parents are 'not entirely' external and objective evaluators, as they have a unique and close relationship with their children. In the case of psychiatric illnesses, both children and parents have reported poorer quality of life than their healthy peers [14, 15]. It is interesting to note that the children's perceptions of themselves and the parents of their children often differ [16, 17]. According to Cummings [18], a comparison of objective and subjective data is essential, and although a weak relationship between objective and subjective indicators can be demonstrated, none can be neglected when examining children [19, 20].

Several studies have examined the extent to which children agree with their parents' perceptions concerning their quality of life [21, 22]. A stronger correlation has been demonstrated concerning the objective areas (*e.g.*, school performance), while a weaker relationship could be detected concerning the child's assessment of the psychological and social situation. Assessing the quality of life of a child can also be influenced by examining the similarities between the evaluation of the parent and the child among both healthy or chronically ill children [23]. Jozefiak *et al.* [22] reported that in the case of healthy children, parents perceive a much more positive status concerning the child's quality of life in almost all areas (except family and friendships) than the children themselves. Hwang *et al.* [24] found that chronically ill adolescents rated their quality of life less poorly than their parents. The reason for this can be that they do not have as much insight into their problems as their parents, so they do not always experience their illness as critical.

Therefore, quality of life is a key factor in gaining a better understanding and more effective treatment concerning people with chronic illnesses. Pediatric health practice also increasingly recognises the importance of integrating illness-specific health-related quality of life (HRQoL) testing into an increasingly holistic approach to disease management [25]. For T1DM, in order to achieve optimal glycemic control, children face serious challenges in their daily lives: having at least 1500 insulin injections within a year, blood glucose measurement with 1000 finger sticks, absence from school of at least 7-15 days due to clinical follow-up examinations, regular contact with the care team, constant self-discipline, and self-control over adherence to the diet. These aspects raise the question of how the requirements of appropriate metabolic regulation and diabetes management, which set strict rules, can affect the quality of life of T1DM children and their parents [26]. Several studies have shown that adolescents with chronic diseases such as kidney disease [27, 28], epilepsy [29], obesity [29], rheumatoid arthritis [30], and sickle cell anaemia [30] have a lower quality of life compared to their healthy peers. This finding is also consistent with the results of research examining the quality of life of adolescents with T1DM [31 - 33].

However, the results of research examining the relationship between quality of life, diabetes management, and metabolic control are ambivalent [26, 34]. Some of the research examining the relationship between quality of life and glycemic control suggests that a higher quality of life score is associated with better glycemic control among adolescents with T1DM. Therefore, in their view, achieving and maintaining an adequate quality of life should be considered as important as achieving optimal metabolic control [35 - 37]. In contrast, others have not demonstrated this (the authors say their results were strongly influenced by the small size of the experimental groups) [38, 39]. In their comprehensive international research, Hoey *et al.* (Hvidore Study Group) [40] investigated the quality of life of more than 2000 adolescents and their parents in 21 pediatric diabetes centres in 17 countries in Japan, North America, and Europe. Their results confirm that, concerning the evaluation of the parents, lower HbA1c levels were associated with significantly better quality of life, higher satisfaction with life, less diabetes-specific concerns, and a lower rate of perception of the negative impact of diabetes on the family. Regarding gender, girls were found to report more concerns, lower levels of satisfaction with life, and poorer quality of life than boys. Parents' perceptions related to the negative impact of diabetes on their daily lives are becoming more positive as the children become older.

Vanelli *et al.* [26] examined the relationships between quality of life, satisfaction with life, and metabolic control among 153 adolescents and their parents. They concluded that there was no significant difference in assessing the impact of diabetes on daily life between boys and girls, which was not affected by age or duration of diabetes but by HbA1c level. Girls reported more significant diabetes-specific concerns than boys. Better glycemic control (lower HbA1c) was associated with better self-rated health, less diabetes-specific anxiety, higher satisfaction with life, and a lower perception of the negative impact of diabetes on family life. Girls rated their health more negatively than boys, and the self-rated health status was lower among girls. Perceptions of the negative impact of diabetes on the daily lives of families have decreased with age.

In their cross-section study, Graue *et al.* [41] examined the diabetes-related quality of life, well-being, and various concerns and satisfaction with life in terms of demographic and clinical variables among 130 adolescents with T1DM and then compared the results with a healthy control group. Their findings showed that adolescents with T1DM reported significantly lower overall quality of life than their healthy peers. Their results concluded that overall health-related quality of life significantly correlated with age and gender, while it did not have a relationship with HbA1c and other diabetes-specific clinical variables. Adolescents with T1DM reported significantly lower overall health than the healthy control group.

In their cross-sectional study, AlBuhairan *et al.* [42] also mapped health-related quality of life and its impact on the family, involving 315 adolescents with T1DM and their parents. Parents' overall average evaluation of their child's quality of life was significantly lower than adolescents' self-evaluation. Adolescents and parents also had the lowest ratings on the Concerns subscale, meaning they were less characterised by diabetes-related anxiety. Female gender and older age were considered predictors of lower quality of life. In terms of the results of the PedsQL Family Impact Module, the lowest score was obtained on the subscale measuring emotional functions. Overall, they concluded that age and gender might be explanatory factors concerning the differences in the quality of life of adolescents with T1DM.

According to Naughton *et al.* [43], quality of life decreases with age among girls and increases with age among boys. The results of Terrason *et al.* [44] also show a relationship between gender and quality of life but found no relationship between age and quality of life. This may be because, in the general population, girls experience a higher prevalence of depressive symptoms during adolescence than boys [45]. Also, ambivalent research results can be found regarding the relationship between quality of life and the duration of T1DM. Parkerson [46] found no correlation, but Dasbach *et al.* [47] reported a better quality of life for a shorter period after diagnosis.

Gender differences also have been demonstrated in the clinical context of diabetes [40]. Girls enter puberty earlier than boys and, for several reasons, have poorer glycemic control [48]. This may also include reduced adherence to various aspects of the treatment regimen and decreased insulin sensitivity of the peripheral tissues [39, 49]. These differences in metabolic control can affect the quality of life in different ways in terms of gender [40]. Diabetes-specific quality of life research among children and adolescents with T1DM conclusively demonstrates that quality of life is better with better glycemic control and among men [50], younger patients, and higher socioeconomic status [51]. To sum up, these results highlight the role of the efforts to map mental health determinants among children and adolescents with T1DM and their parents to support patients in achieving better metabolic control.

Self-Rated Health In Diabetes

The basis of self-rated health is the perception of an individual's own state of health. This assessment is usually based on the conscious or unconscious comparison with peers living in the individual's environment [52]. It is a continuous dimension with a 'very favourable' perceived state at one end and a 'very unfavourable' perceived state at the other end. Self-rated health status is a significant indicator of well-being, as health status in childhood and adolescence is a major predictor of quality of health in adulthood [53]. Mapping the incidence of subjective health complaints and their impact on their lives is becoming increasingly important among adolescents [54]. This includes several bodily symptoms that can range from transient malaise to clinical conditions that prevent the person's daily functioning and require long-term medical supervision [55]. Although we distinguish between 'somatic' (*e.g.* pain) and 'psychic' (*e.g.* bad mood) health complaints, it is important to emphasise that these cannot be identified as the cause of the symptom. For example, fatigue can be a symptom of depression, while inflammatory diseases can cause withdrawal and depressed mood [56]. Due to the heterogeneity of individual experiences and reactions, it is difficult to set a generally applicable watermark or a clinical limit concerning subjective health complaints [55]. Although adolescents regularly report such symptoms, in part due to increased self-observation caused by physical and mental transformation, they do not generally interfere with their daily functioning, nor do they indicate emotional disturbance [56]. It is important to emphasise that the majority of adults reporting common physical complaints in adulthood were characterised by common physical symptoms previously in adolescence. A chronic health condition that requires regular medical checkups, such as diabetes, can be a source of stress and limit the adolescent's ability to meet his or her increased need for autonomy. However, according to the study of Aszmann *et al.* [8], the difference between the frequency of physical and mental symptoms in chronic and healthy adolescents was not significant.

Satisfaction is a global, quantifiable assessment of subjective well-being. It covers the dimension of health and also expresses the individual's overall quality of life. It has a fairly high temporal stability, and since it has a more cognitive nature, it is less affected by current emotions and moods. Adolescents' satisfaction with life is significantly determined by life experiences and relationships, especially in the context of the family [57]. Life satisfaction and school-related factors (*e.g.* academic achievement, relationship with classmates and teachers, or bullying) interact [58].

Illness Representations

Examining illness representations, the personal meaning of the disease [59], and adherence can enrich psychological work with children with chronic illness in several ways. They determine the patient's implicit beliefs about their disease when assessing symptoms and the coping strategies they choose [60]. Studies in people with diabetes and other chronic diseases have shown that illness representations play a significant role in explaining the diversity of patients' attitudes and coping with their disease [60]. Illness representations, also known as illness schemes, are subjective illness models, cognitive representations, cognitions of the illness, and illness-related thoughts. Any mental activity that is related to how the patient thinks about their illness, how they experience it, how they display it in their inner world of experience, *i.e.* the patient's cognitive representations of illness, is an extension of the schematic theory of cognitive social psychology [61]. An individual's ideas, beliefs, and explanations about their illness and thus the subjective theories of illness are determining factors in the regulation of illness behaviour and self-management. A patient's self-management attempts are shaped mainly by their subjective illness model, even if their beliefs do not fit the medical model of illness. Illness representations influence the perception of symptoms, psychic reactions to the disease, the time of asking for help, the perception of control over the treatment, and the whole process of adherence to treatment [62]. Following Leventhal, the development of lay theories of illness aims to give meaning to the patients' often diffuse, alien, and anxious feelings associated with their altered condition. Five dimensions of illness representations can be distinguished based on which patients structure their experiences of their illness. These dimensions are the identity (name, label and characteristic symptoms of the disease), the causes of the disease (factors responsible for the development of the disease), time course (beliefs about the duration of the disease), consequences (expectations about the physical, psychological, social and economic consequences of the patient's daily life), and controllability (whether the course and symptoms of the disease can be influenced) [63].

Research focusing on adolescents with diabetes confirms the importance of illness representations in mapping the medical and psychological characteristics of this group of patients [61]. Studies of adolescents with T1DM have found an association between children's beliefs about diabetes and their emotional well-being (both cross-sectional and longitudinal studies). However, neither the perceived long-term efficacy of medical therapy nor the perceived severity of diabetes predicted children's self-management of adherence or the degree of distress experienced concerning diabetes [61, 64]. Furthermore, the association between illness representations and blood glucose and insulin treatment could not have been confirmed [64]. Nevertheless, no association was found between illness perception and HbA1c, which may be due to the fact that illness perception affects metabolic control but only among girls [65].

Concerning emotional well-being, research among children with T1DM has demonstrated that illness representations related to the consequences of diabetes are associated with symptoms of anxiety and depression [64, 66]. However, perceived personal control [66] and perceived efficacy of medical therapy are predictors of subjective well-being in children with diabetes [64, 67, 68].

FACTORS INFLUENCING ADHERENCE

T1DM self-management includes that the patient actively monitors and responds to the changing environmental and biological conditions and adapts to the different treatment tasks in order to maintain appropriate metabolic control and reduce the likelihood of complications [69].

Self-management involves:

- regular glucose monitoring (blood or urine) to achieve appropriate metabolic control and avoid long-term complications,
- following a proper diet, especially adjusting your carbohydrate intake,
- insulin treatment
- regular exercise and
- participation in regular medical check-ups [70].

Diabetes-specific adherence can be conceptualised as the active, voluntary collaboration of an individual in treating a disease, following a mutually agreed treatment procedure, and sharing responsibilities between the patient and the health care provider [71]. Hentinen [72] defines therapeutic collaboration as an active, responsible, and flexible self-management process in which the patient, instead of rigidly following the prescribed treatment instructions, actively seeks to achieve adequate health in close collaboration with the healing team. Another important concept is 'unintentional nonadherence', which occurs when the patients think they are cooperating with the recommended treatment regimen but are actually unable to do so due to a lack of knowledge or skill.

Concerning diabetes-specific adherence, it is crucial that the degree of cooperation with all components of the treatment regimen should be assessed independently (*i.e.* blood glucose self-monitoring, insulin therapy, diet, physical activity, and other self-management tasks) rather than just to assess the level of cooperation concerning one single treatment regimen. This is important as it is increasingly proven that there is only a slight correlation between adherence and individual treatment tasks, suggesting that adherence is not a one-dimensional construct [69, 73]. This finding has been reported in both type 1 and type 2 diabetes [73]. Furthermore, various associations between adherence and adequate metabolic control have been demonstrated among people with diabetes [74].

Adherence is a multidimensional phenomenon defined by the interaction of five groups of factors (dimensions), of which patient-related factors are only one dimensional. The general belief that the patient is solely responsible for the treatment is deceptive. It often reflects a misunderstanding of how other factors affect people's behaviour and ability to interact with their therapies. The five dimensions of adherence are the health care system, socioeconomic, therapy-related, condition-related, and individual factors [75 - 79].

Perceptions of the need for medication are significantly influenced by the severity of symptoms, expectations, experience, and illness representations [80]. Concerns about medication usually arise from beliefs about side effects, experiences of previous lifestyles, and abstract concerns about long-term side effects and addiction. Motivation for therapeutic collaboration is outstandingly determined by the individual's personal belief in the importance of cooperating with treatment, as well as a belief in its value and a belief in one's abilities to follow treatment instructions [80]. Developing intrinsic motivation by increasing the perceived importance of adherence and strengthening the patient's self-confidence in self-management skills is a behavioural intervention goal that is important to use in parallel with biomedical goals if the goal is to improve adherence sustainably and effectively.

Adherence In Adolescence

The health care needs of children and adolescents change with their growth and development, regardless of the type of diabetes. From a psychological point of view, it is also essential to consider the changes that occur during the different stages of developmental psychology in the management of diabetes. Currently, several guidelines are in use for the management of children and adolescents with type 1 diabetes:

- National Evidence-Based Clinical Care Guidelines for Type 1 Diabetes in Children, Adolescents and Adults (2012) Australian Paediatric Endocrine Group and the Australian Diabetes Society.
- Standards of medical care in diabetes: special considerations for children and adolescents (ADA 2013)
- SIGN (2010)

- Nice Guidelines
- Clinical Practice Consensus Guidelines (2009) Compendium International Society for Paediatric and Adolescent Diabetes.

The comprehensive goal of these guides is to provide personalised education and management to achieve the following goals:

- The establishment of an accurate diagnosis.
- The prevention or delay of complications, including short-term complications, *e.g.*, hypoglycemia and ketoacidosis. Long-term complications before puberty are rare, but screening for complications usually begins around the age of 10 or even earlier.
- To create a balanced and nutritious diet following the child's stage of growth and development.
- The acceptance of diabetes by the child and family.
- To support the child in gradually taking over the tasks of self-management.
- To develop a holistic and individual health plan program that includes support for mental and sexual health, responsible contraception, and planned pregnancy, tailored to age and current stage of development.
- The measurement of HbA1c levels every 3rd-4th months, with an individually determined target range, but in general, it is important to set the values below 7.5% without hypoglycaemia.
- The admission to acute care if necessary.
- Start screening for microvascular complications, usually between the ages of 10 and 12.
- A smooth transition to adult care that requires collaboration between child and adult care services.
- Health centers and schools/educational institutions involved in diabetes care should cooperate to ensure that the child's self-management can function smoothly at school.
- School staff should have adequate knowledge of, for example, the treatment of hypoglycaemia and, if necessary, have emergency telephone numbers.

It is crucial to highlight that almost all guidelines emphasise the essential role of multidisciplinary teamwork to achieve optimal diabetes management. In the daily care of patients, the collaborative work of the team members (pediatric endocrinologist, dietitian, psychologist, physiotherapist, social professionals) is unquestionable. Furthermore, it is essential to emphasise the importance of involving the child in the treatment of diabetes according to his/her abilities and possibilities following his/her age. The extent of the child's participation and involvement in diabetes management gradually increases with maturity and the development of fine motor, problem-solving, and coping skills. The recommended metabolic values and HbA1c levels should be in line with those expected for age and developmental stage and should be reviewed regularly as the child develops.

Concerning the stages of life, due to their prominent biological and psychological significance, the challenges of adolescence and the management of type-1 diabetes in adolescence should also be highlighted. Keeping glycemic control in adolescents with diabetes in the optimal range is highly dependent on the degree of adherence [68], and the role of appropriate self-management and adequate glycemic control in preventing long-term complications have been demonstrated [82]. However, proper self-management is often not implemented in this age [83, 84]. Although chronic diabetes-specific complications are rare in adolescents, their pathogenesis begins soon after diagnosis is made, and the acceleration of the pathological condition is observed during the puberty phase [85]. Among adolescents diagnosed with diabetes, glycemic control aims to keep HbA1c levels below 58 mmol/mol (7.5%) and minimise hypoglycaemic episodes [86].

Several studies confirm that the majority of adolescents with diabetes do not adhere to an adequate therapeutic regimen [87 - 89]. The rate of nonadherence for glucose monitoring ranges from 30% to 80% [90]. According to Cox [91] and Petitti [92], young people with HbA1c levels outside the target range in the US would urgently need to develop effective treatment strategies, as the significant reduction in therapeutic adherence and metabolic control at this stage of developmental psychology requires clinical attention [93 - 95]. It is important to emphasise that new treatment methods and approaches are needed given the complexity of T1DM care in adolescence [96]. Inadequate therapeutic adherence, through suboptimal glycemic control and diabetes-specific complications, can have serious clinical consequences [97], can be an acute complication of diabetic ketoacidotic coma, and can increase morbidity and associated medical conditions. Furthermore, it can increase the number of hospitalisations and clinical visits, leading to higher overall mortality [98, 99]. It is also associated with several mental disorders, such as depression [88, 99], eating disorders [100], and pathological anxiety from hypoglycemia [101]. Therefore, children with inadequate glycemic control may face extremely severe consequences in terms of quality of life, family relationship dynamics and functioning, and financial burdens [102, 103].

Factors Influencing Diabetes-Specific Adherence

Factors influencing diabetes-specific adherence can be categorised into four groups:

- the characteristics of the illness and its treatment,
- intrapersonal factors,
- interpersonal factors,
- environmental factors.

The Characteristics Of The Illness And The Related Factors

Several characteristics of diabetes and its treatment have also been associated with therapeutic cooperation, including the complexity of the treatment and the duration of the illness. Regarding the complexity of treatment in general, research has concluded that the more complex the treatment regimen, the less likely the patient is to follow it. Indicators of the complexity of the treatment are the frequency of self-management behaviours (*i.e.* how many times a patient needs to perform one behaviour per day). Paes *et al.* [104] examined the prevalence of inadequate therapeutic cooperation among 91 diabetic individuals and its correlations with the frequency of drug dosage. Their results showed better adherence in individuals with a lower frequency of medication (once per day) than more frequent (three times a day) medication use. Dailey *et al.* [105] stated a similar conclusion, with patients taking a single drug showing better short- and long-term adherent behaviour than those taking two or more drugs simultaneously.

According to the literature, a negative relationship can be found between the duration of the illness and adherence. Thus, the longer the duration of diabetes, the less likely the patient will show adequate therapeutic cooperation. Glasgow *et al.* [67] examined adherent behaviour in various diabetes-specific areas (insulin treatment, blood glucose measurement, diet, physical activity) and their association with glycemic control among 93 individuals diagnosed with T1DM (mean age 28 years). Concerning the treatment areas, a significant association was found between physical activity and the duration of diabetes. Patients who have been living with diabetes for less than ten years have reported higher levels of physical activity (exercise several times a week) than those who have had diabetes for more than ten years. Patients with a long medical history also reported a higher proportion of foods that were not suitable for their diet, meaning that they were less able to follow their diet. Jaros-Chobot *et al.* [106] investigated 183 American and 80 Polish children with T1DM (mean age 13 years) to map the associations between adherence and glycemic control. Their results showed a correlation between the duration of the illness and insulin treatment. Children with a long medical history were more likely to forget their insulin injections than those with a shorter duration of diabetes.

Intrapersonal Factors

Several important factors were found to be associated with adherence. These were age, gender, self-esteem, self-efficacy, stress, and comorbid depressive disorder. According to Glasgow *et al.* [69], there is an association between age and physical activity in patients with T1DM. Patients over 25 years of age reported a lower frequency of physical activity than the younger age group. No association was found between age and other diabetes-specific treatment areas. Weissberg-Benchell *et al.* [107] examined these associations among 144 adolescents with T1DM. They concluded an association between age and insulin treatment, meaning that older people were more likely to receive inadequate insulin treatment than younger children. Anderson *et al.* [108] studied 89 children (age between 10–15 years) to map the association between adherence and glycemic control. Their results showed that younger children reported a higher frequency of blood glucose self-monitoring than older ones.

Research has also found a relationship between gender and adherence. Glasgow *et al.* [69] found that men reported a higher frequency of physical activity than women while consuming more calories, following a proper diet, and showing a lower level of therapeutic cooperation with diet.

There was also a relationship between self-esteem and adherent behaviour in terms of self-management. Patients with T1DM with higher levels of self-esteem reported a higher frequency of physical activity and adequate insulin therapy [109]. Murphy-Bennett *et al.* [110] found that lower self-esteem was associated with less frequent self-monitoring of blood glucose levels among adolescents with T1DM.

The role of self-efficacy has also been studied in adhering to prescribed treatment for diabetes. According to the results of Plotnikoff *et al.* [111], diabetes-specific self-efficacy beliefs proved to be the strongest predictors of physical activity. Senecal *et al.* [112] reported that self-efficacy beliefs were strong predictors of adherent behaviour, and both a sense of self-efficacy and appropriate therapeutic cooperation predicted life satisfaction among patients with T1DM. Ott *et al.* [113], in their research with 143 adolescents diagnosed with T1DM, concluded that self-efficacy was a significant predictor of the extent of therapeutic cooperation in diabetes.

The literature confirms that stress and emotional problems are also significantly associated with adherence. A strong association can be found between stress and adherence to a diabetes-specific diet (in terms of the amount and type of diet) among patients with type 1 and type 2 diabetes. Peyrot *et al.* [114] examined the relationship between adherence, diabetes-specific distress, and glycemic control among 57 patients with T1DM and 61 patients with T2DM. Their results showed that the higher the psychosocial distress reported by a patient, the less appropriate is the therapeutic interaction with the prescribed treatment regimen, and the poorer is glycemic control. Mollema *et al.* [115] found that T1DM patients with extreme fear of insulin treatment and blood glucose measurement were characterised by lower levels of therapeutic cooperation and higher levels of emotional distress. Schlundt *et al.* [116] classified T1DM patients into different groups according to the problems they experienced in adhering to the prescribed diet. They found that two groups of patients, emotional eaters and diet-bingers, could have been characterised by negative emotions such as stress and depression and inappropriate adherent behaviour.

The incidence of depression is twice as high among patients with diabetes as in the general population [117]. Patients with comorbid depression are more likely to develop complications [118], have poorer glycemic control [93], and are less cooperative with the therapeutic regimen than those without depression. Comorbid depression is also associated with higher costs [79].

Interpersonal Factors

Several studies have identified the relationship between interpersonal factors and therapeutic collaboration. These investigations mainly focused on the relationship between patients and health professionals and the quality of social support. Adequate communication between patients and healthcare professionals was associated with a higher level of adherence. Among T2DM patients, thc use of oral antidiabetics and monitoring of blood glucose levels were significantly worse in patients who did not consider their communication with their physicians to be satisfactory [79]. As one of the most significant components of social support among children with T1DM, the degree of parental involvement showed a significant correlation with adherence to blood glucose self-monitoring. Children and adolescents with T1DM who experienced higher parental involvement in blood glucose self-monitoring reported a higher level of daily blood glucose self-monitoring [109]. Glasgow *et al.* [69] examined adherence in a sample of adolescents and adults with T1DM, concluding that in both adults and adolescents, illness-specific social support was associated with a higher level of adherence in the areas of insulin treatment and blood glucose measurement. Also, several studies have demonstrated a relationship between a low level of social support and inadequate adherence [116, 119].

Environmental Factors

Two environmental factors, namely high-risk situations and environmental systems, have been associated with inadequate adherence among diabetic youth.

Adherent behaviour must also be maintained in an ever-changing environment (*e.g.* home, school, workplace, *etc.*) with different needs and priorities. As circumstances change, patients face various challenges to maintain proper self-management. Patients are often faced with a choice between caring for diabetes management and paying attention to other life-giving priorities. Situations related to inadequate adherence are called high-risk situations [120].

Schlundt *et al.* [116] developed a taxonomy of high-risk situations that make it difficult for patients to comply with dietary requirements. These situations may be related to consuming too much or too little food or situations that involve difficulties in incorporating the meal into a particular social context or time of day. According to Schlundt *et al.* [120], loneliness, boredom, and interpersonal conflicts, as well as meals at school, social events, or holidays, can all be high-risk situations for adhering to a proper diet. Different categories of high-risk eating situations have also been identified for adults with diabetes that are resistant to temptation, eating out, situations under time pressure, and situations where there is competition between priorities and social events. Other research has shown that environmental barriers can predict adherent behaviour associated with different areas of diabetes self-management [121, 122].

THE PURPOSE OF THE STUDY: THE CONNECTION BETWEEN MENTAL HEALTH, ADHERENCE AND DIABETES

The aim of the research is to investigate the psychological characteristics and adherence of children and adolescents with type 1 diabetes. Based on the Hungarian and international literature, no detailed and complex adherence questionnaire has been developed so far. For this reason, in the pre-research phase, a complex questionnaire with 9 subscales was created following the requirements of the standard test development procedure, which was already used as a measurement tool in the present research. Furthermore, the aim of the research is to determine the effect of mental health factors on adherence so that we can receive information concerning the psychological effects of living with diabetes.

Sample Characteristics

This research is a part of complex research where healthy and chronically ill children are compared by their mental health. In this present study, we only focus on children with T1DM and their mental health following its various areas.

The research was based on the Diabetes Adherence Survey 2017 (hereinafter DAS 2017). The study was carried out partly in Pediatric Psychology and Psychosomatic Unit at the Institute of Pediatrics of the Faculty of Medicine at the

University of Debrecen and partly at some primary and secondary educational institutions in Debrecen.

Experimental Group: Children Diagnosed With Type-1 Diabetes

Children diagnosed with T1DM participating in the research (N=114) received treatment in the Endocrinologocal Ward of the Clinical Centre of the University of Debrecen. The research was carried out at the Pediatric Psychology and Psychosomatic Unit at the Institute of Pediatrics of the Faculty of Medicine at the University of Debrecen during the study (ethical permission ID: 4528A -2016, provided by the regional ethical committee [DE RKEB/IKEB]). The children and their parents participating in the study were informed in written consent about the possibility to participate in the research, its purpose and its voluntary nature.

Table 1. Sociodemographic and illness-related characteristics of children belonging to the experimental and control group.

Sociodemographic Data	Experimental Group	Control Group
Age (in years) (M, SD)	14,13 (2,47)	14,08 (1,99)
Gender (N, %) boy girl	62 (47,69) 68 (52,31)	76 (42,93) 101 (57,07)
Family structure One parent (M, SD) Two parents (M, SD)	44 (33,85) 86 (66,15)	49 (27,69) 128 (72,31)
Relatioship with mother Biological mother (M, SD) Adoptive mother (M, SD) Foster mother (M, SD) N/A (M, SD)	99 (76,15) 1 (0,007) 0 (0) 30 (23,297)	117 (66,1) 2 (1,1) 5 (2,9) 53 (29,9)
Mother's educational level Tertiary level Secondary level Primary level	30 (23,2) 55 (42,3) 15 (11,5)	55 (31,1) 117 (66,1) 5 (2,8)
Data related to T1DM		
Duration (in years) ***(M, SD)***	6,72 (3,78)	-
HbA_{1C}	8,32 (1,53)	-
Type of insulin therapy MDI CSI	58 (44,61) 72 (55,39)	- -

The youngest child was 9 years old, and the oldest was 18 years old, and the mean age of the sample was 14.17 years (SD=2.18). Data collection was carried out between September 2017 and May 2018. Regarding the gender distribution, 52.6% of the sample were boys (N=60), and 47.4% were girls (N=54). 67.5% of the participants live in an intact family, 21.9% live with only one parent (and/or its new partner), and 8.8% live with foster parents. 16.7% have no siblings, while 83.3% have at least one sibling. The mean age of children at the occurrence of symptoms of T1DM was 8.18 years (SD=3.9). By the time of our research, the mean duration since diagnosis was 7.1 years (SD=3.8). 43.9% of children use an insulin pump, while 56.1% receive conventional insulin therapy. For children receiving insulin pump therapy, the mean duration of using insulin pump was 3.7 years (SD=2.3). The sociodemographic and diabetes-specific characteristics of the sample are shown in Table **2**.

Table 2. The items and factor weights of the Diabetes Adherence Questionnaire (DAQ) (N=114).

Items	Factor Weight[1]
1. Factor: Self-management *The factor expresses the understanding, organisation and execution of the treatment regulations of diabetes and their careful management according to the treatment protocol*	
I follow the suggestions of the schedule.	0,385
I eat an appropriate amount of food.	0,485
I eat at an appropriate time (breakfast, snack, lunch *etc.*).	0,352
I take an appropriate amount of insulin.	0,234
I take insulin at the appropriate time.	0,364
When taking insulin, I take my blood pressure into account.	0,425
I measure my blood glucose level in all circumstances.	0,412
I record my blood glucose level regularly.	0,236
I do my best to control and keep diabetes in line.	0,352
During meals, I eat an appropriate amount of food following my diet	0,354
2. Factor: Emotional feedback (emotional reactions associated with blood sugar level measurement) *The factor expresses diabetes-specific emotional reactions (e.g,* related to blood glucose levels).	
I prefer to let my family members communicate with my environment about diabetes.	0,42
I prefer to communicate with my environment about diabetes on my own.	0,385
It evokes guilt when my blood sugar is high.	0,541
It evokes fear when my blood sugar is low.	0,428
I can control my fears related to diabetes.	0,395
I become afraid because of my high level of blood glucose	0,394

(Table 2) cont.....

Items	Factor Weight[1]
I become afraid because of my low level of blood glucose	0,457
3. Factor: Social support (parents and family) *The factor expresses diabetes-specific family support*	
My family encourages me to keep my diet.	0,299
My family remembers me to keep my diet.	0,318
My family members control my appropriate consumption of meals	0,361
My family members prepare dietary meals for me.	0,428
My family members commend me when I keep my diet and eat healthily.	0,379
My family remembers me when I eat inappropriate food.	0,364
My family commends me for doing physical activity.	0,502
My family members encourage me to pursue sport.	0,428
My family is glad when I appropriately treat diabetes	0,395
My family commends me when following the schedule.	0,382
My family members help me in changing the treatment on the basis of my blood glucose.	0,425
4. Factor: Social support (peer relationships) *The factor expresses the attitude of the child with diabetes towards peer relationships.*	
The relationship with friends is more important than treating diabetes.	0,348
I am discriminated in my friendships due to diabetes.	0,531
I avoid my peers becoming aware of my diabetes if possible.	0,381
It is disturbing when my friends ask about diabetes.	0,297
Most of my friends are not diagnosed with diabetes; thus, I feel different.	0,405
5. factor: Denial of the disease *The factor involves denying the fact of diabetes and its conscious or unconscious ignoration.*	
Nothing wrong can happen to me if I do not follow the treatment schedule.	0,295
I try to forget diabetes.	0,308
Sometimes I forget the treatment of diabetes.	0,328
6. Factor: Positive adherence *The factor expresses the positive effect of adherence to treatment regulations and regular exercise and sports on physical and mental well-being.*	
I do enough physical activity.	0,451
The more I follow my treatment schedule, the better I feel.	0,392
7. Factor: Negative adherence (the burden of the treatment) *The factor expresses the negative consequences of living with diabetes on everyday life (e.ge.g., social relationships, leisure, and school activities).*	
My treatment causes inconvenience.	0,364
My treatment requires too much time and work.	0,421

(Table 2) cont.....

Items	Factor Weight[1]
Diabetes limits my friendships.	0,381
Diabetes strongly limits my lifestyle.	0,356
The appropriate treatment of diabetes requires a huge amount of time.	0,348
My diet significantly limits my lifestyle.	0,360
Diabetes constrains me from pursuing sport.	0,295
Diabetes limits my leisure activities.	0,278
Diabetes limits my school activities.	0,347
8. Factor: Social support (medical team) *The factor expresses the child's relationship and attitude toward the medical team and the perception of their behaviour.*	
I trust doctors.	0,395
Doctors are too busy to discuss diabetes with me.	0,402
Doctors understand that diabetes hinders me from doing things that are important for me.	0,414
Doctors are friendly and can easily talk to me.	0,415
I trust nurses.	0,395
9. Factor: Vision (worries) *The factor examines concerns and fears about the long-term negative consequences of diabetes (e.ge.g. in the areas of marriage, childbearing, further education, and employment).*	
I am afraid whether I will marry due to diabetes.	0,361
I am afraid whether I will have any child due to diabetes.	0,328
I am afraid that it will be harder to find a job due to diabetes.	0,297
I am afraid that I cannot finish my studies due to diabetes.	0,368
I am afraid that diabetes influences my physical appearance.	0,328
I am afraid that I will have medical complications due to diabetes.	0,401

Control Group

Healthy children and adolescents participating in the study (N=233) were randomly recruited. Similar to the patient group, the members of the control group were informed in a written consent about their participation in the research, its methods, its purpose and its voluntary nature. The examined children were recruited from the 9th to 12th grades of a vocational high school and from the 5th to 8th grades children of a primary school in Debrecen. The exclusion criterion was the presence of any chronic disease. Thus, a total of 56 children were excluded. The mean age of the sample was 14.08 years (SD=1.99). The sociodemographic and diabetes-specific characteristics of the sample are shown in Table **1**.

Applied Tools

Demographic Questions

First, a demographic questionnaire was used to determine the following information related to the child:

- gender
- age
- type of residence (capital/county seat/big city/small town/village/farm)
- the highest educational level of the mother/foster mother
- the highest educational level of the father/foster father
- structural change in the family and its nature, if any ('yes, my parents are divorced and my mother is raising me alone'; 'yes, my parents are divorced and my father is raising me alone'; 'yes, my parents are divorced, my mother is raising me and has a new relationship; 'yes, my parents are divorced, my father is raising me and he has a new relationship'; 'I don't live with my parents'; 'other')
- number of siblings (both biological and foster)
- birth order
- subjective financial situation

The Creation of the Diabetes-Specific Adherence Questionnaire

As the first step of our research, we developed an adherence questionnaire to explore the therapeutic collaboration of children and adolescents with type 1 diabetes and the factors influencing it. Reviewing the literature, we found ten relevant international questionnaires that examined the attitudes, adaptation, and therapeutic interactions of children and adolescents with T1DM to diabetes, treatment and lifestyle changes. However, these questionnaires only cover certain sub-areas of adherence. The researchers' approach is mostly similar, but their approaches are different in terms of content. Thus, the items of the ten questionnaires were categorised into content categories by metaanalysis. Then their proper coherence was examined, overlapping contents were omitted and merged, thus avoiding the loss of the content categories. After sorting the content categories of the previously applied questionnaires, independent juries' finalisation of the content categories was done. Thus, the first version of the questionnaire was created by translating ten English questionnaires and then translating them back into the original language, which contained 167 statements in this form. The questionnaires used were the following:

1. The Diabetes Family Behavior Scale (DFBS) [123]
2. Diabetes Family Behavior Checklist (DFBC) [124]
3. Diabetes Family Responsibility Questionnaire (DFRQ) [125]
4. Diabetes Social Support Questionnaire – Family Version (DSSQ) [126]
5. Blood Glucose Monitoring Communication Questionnaire (BGMC) [127]
6. Self Care Inventory – Revised Version (SCI-R) [128]
7. Diabetes Quality of Life for Youth [129]
8. Short-Form Self-Efficacy Item (SEI) [130]
9. Diabetes Family Conflict Scale – Revised Version (DFCS) [131]
10. Problem Recognition and Illness Self-Management (PRISM) [132]

The items were screened with ten professionals (psychologists) and a group of ten pediatric patients diagnosed with T1DM, resulting in the identification of nine subscales, but for later categorisation, factor analysis was performed to reduce the margin of error. Our questionnaire containing 167 statements was pre-tested by 20 individuals, based on which we found it appropriate to assess diabetes-specific adherence in children.

The Factor Analysis Of The Questionnaire

Factor analysis was used to detect the most significant factors of the questionnaire. To determine factor selection (extraction), we tried to maximise the variance of the factors. We examined how many independent factors the 167 items can be separated in the total variance of the items to determine the number of scales. Thus, varimax rotation was used to generate factors from which the most interpretable ones were then selected. Among the factors formed by Varimax rotation, the 9-factor version proved to be well-understandable. Then items with an extraction value above 0.1 were left. Thus, a well-interpretable factor structure with a suitable factor weight was obtained.

The reliability of the whole questionnaire was very high (Cronbach α=0.739) since a Cronbach α value above 0.6 is already considered acceptable. However, completing the questionnaire proved to be very long as it took 50-60 minutes per person. Thus, due to the children's attentional limitations and age characteristics, we designed a shorter but substantively identical questionnaire. This was carried out by combining the methods of item-item correlations, reliability testing, and content analysis. The new abbreviated complex questionnaire was tested in 114 patients. Factor analysis was used to examine the subscales sampled in the questionnaire, assuming that a pattern will follow the previously detected patterns of the long version of the questionnaire. We used varimax rotation using the maximum likelihood method to measure this, which explains the variables in 71.4%. As a result of the above-mentioned factor analysis, a 9-factor adherence questionnaire was obtained. The factors are shown in detail in Table **3**.

Table 3. The reliability of the scales of the Diabetes Adherence Questionnaire (DAQ) (N=114).

Scales	Cronbach-α
1. Self-management (SM)	0,844
2. Emotional feedback (EF)	0,623
3. Social support (family) (SSF)	0,882
4. Social support (peers) (SSP)	0,674
5. Denial of the disease (DD)	0,714
6. Positive adherence (PA)	0,745
7. Negative adherence (NA)	0,839
8. Social support (medical team) (SSM)	0,673
9. Vision (V)	0,841

Based on the results of the factor analysis, a similar distribution could be detected concerning the groups of questions. Thus, similar to the original, we could establish nine subscales. The reliability of the abbreviated questionnaire is very high (Cronbach-α=0.881). The reliability of the subscales is illustrated in Table **4**.

Table 4. Gender differences in the subscales of Diabetes Adherence Questionnaire (Source: DAS, N=11).

Subscales	Boys (N=60)		Girls (N=54)		p
	M	SD	M	SD	
Self-management	40	10	38	11,8	0,350
Emotional feedback	16,5	3,6	15,4	4,9	0,003
Social support (parents and family)	47,2	13,9	47	15,2	0,787
Social support (peer relationships)	18,3	3,1	18,4	5,1	0,188
Denial of the disease	5,6	2,1	5	2	0,275
Positive adherence	7,9	2	7,8	2,7	0,584
Negative adherence	16,2	7,3	13,8	6	0,197
Social support (medical team)	21,9	4,2	21,6	5,9	0,461
Vision	11,8	5,8	8,8	4,4	0,002
Total	184,3	34,5	175,9	48,1	0,491

Children Depression Inventory (CDI) [133]

CDI is a 27-item questionnaire used to measure the level of depression (frequent mood swings, self-esteem, and social behaviour problems) among children aged 7 to 18 years, with three choices per question (0,1,2). Of the responses, "0" indicates no symptoms, "1" indicates mild symptoms, and "2" indicates a marked presence of symptoms in the past two weeks. The maximum available score is 54. In the present study, we worked with two values, distinguishing between a group at risk of depression (13–15 points) and a group with clinical depression ($\geq$ 16 points). The questionnaire examines sadness, anhedonia, self-hate, indecision, suicidal thoughts, interpersonal relationships and feelings of being unloveable. The reliability of the questionnaire (Cronbach α=0.92) indicates that the internal consistency was found to be adequate.

World Health Organization Well-Being Index (WBI-5) [134]

WHO General Well-Being Index aims to provide information on the general well-being of individuals over the past two weeks. It is one of the most commonly used questionnaires to assess general subjective well-being. The questionnaire measures well-being through five statements. It is a short and quick-to-measure tool of positive well-being that can be characterised by reliable psychometric characteristics. In the Hungarian version [135], statements should be answered on a 4-point Likert scale (0-3, not typical at all / barely typical / typical / completely typical). WBI-5 is an appropriate tool for examining emotional problems over the past 14 days. It has also been used in large-sample representative research, such as the Health Behavior in School-aged Children.

Self-rated Health (SRH) [135]

The self-rated health status was examined with the question, 'How would you rate your health status compared to people of similar age?' Children had to rate the question on a 4-point Likert scale, where 1 meant 'poor', 2 meant 'appropriate', 3 meant 'good', and 4 meant 'excellent'. The question has been shown to be reliable in both adult and adolescent populations [136].

Psychological Mood and Somatic Symptoms [136]

In addition, we examined the frequency of nine subjective health complaints as psychological and somatic symptoms. Children had to rate the questions on a 5-point Likert scale (1-5, almost never / rarely / occasionally / often / almost always). The prevalence of the following symptoms was examined: headache, diarrhea due to nervousness, back and/or low back pain, irritability, nervousness, sleeping difficulties, and fatigue.

Satisfaction with Life, SWL-present (SWL-p) SWL-future (SWL-f), Cantril-ladder [137]; Life Evalution Index [138]

Satisfaction with life (SWL) was measured using the Life Evaluation Index [138], which was developed based on the eleven-point scale of Cantril [137]. The Cantril ladder is a visual analogue scale. Thus, the marked grade shows where the respondent places itself between the two endpoints of the dimension. Children were asked to rate themselves on an 11-point ladder to assess their degree of satisfaction with both their present and future life situation five years later. Step 10 of the ladder is the highest; step 0 denotes the lowest degree of satisfaction with life for both the present and future life situations. The Cronbach's alpha value of the index is 0.91. Gallup [139] formed three independent groups based on scores on the index:

1. "Thriving": they are characterised by strong and consistent mental well-being. They evaluate both their current life situation and the next five years positively. Significantly fewer health problems, worries, stress, sadness, and anger are reported; however, they are characterised by higher levels of happiness. In their case, the score for the present evaluation is ≥ 7 and for the future ≥ 8.
2. "Struggling": in their case, mental well-being is inconsistent. They are moderately satisfied with their current or future life situation. Compared to the thriving group, they report higher levels of daily stress and financial anxiety and are twice as likely to become sick.
3. "Suffering": this group is at high risk for developing mental disorders. They are slightly satisfied with their current life situation, and their assessment for the next five years is also very low. More frequent somatic complaints and a higher burden of illness, stress, anger and sadness is reported. Present and future evaluation scores are ≤ 4.

Pediatric Quality of Life Inventory, PedsQL Measurement Model [139, 140]

The Pediatric Quality of Life Inventory is a multidimensional measurement tool suitable for examining the health-related quality of life (HRQOL) of healthy and ill young people with chronic and acute illnesses between the 2-18. This tool measures children's quality of life with various chronic diseases (diabetes, obesity, oncological, cardiological, rheumatological, neuromuscular problems, *etc.*) and healthy children as well. According to the theoretical background of PEDsQL, children experience the impact of health-related quality of life on the dimensions of health and well-being related to the disease and treatment.

The PEDsQL 4.0 General Questionnaire scale consists of 23 questions and includes the following subscales: physical functioning (eight items), emotional functioning (five items), social functioning (peer relationships, social activities) (five items), and school functioning (five items) [140, 141].

Satisfaction with Life Scale (SWLS) [141]

The Satisfaction with Life Questionnaire is a frequently used tool to examine satisfaction with life as an indicator of subjective well-being. The Hungarian version of the questionnaire was adapted and validated for the Hungarian population and is proved to be reliable (Cronbach α=0.885) [142]. The results of the questionnaire are related to mental health indicators and may be one of the predictors of future health behaviour. It is also often used to assess the mental health of the populations with some type of chronic physical or mental illness.

Strengths and Difficulties Questionnaire (SDQ) [142]

The Strengths and Difficulties Questionnaire (SDQ) allows us to easily and quickly screen behavioural problems and mental disorders. The widely used tool developed by Goodman [142] makes it possible to map the difficulties and strengths of children aged 4–17 years. The Hungarian SDQ questionnaire consists of 25 items, grouped into five scales, each containing five items. The scales measure emotional, behavioural, hyperactivity and peer relationship problems and prosocial behaviour. The five-scale questionnaire can be used from the age of 4 by interviewing parents and teachers. It can also be used in self-completion form from the age of 11, which allows the problems to be examined from several perspectives. The questionnaire allows us the quick screening of problematic cases. The Emotional symptoms subscale includes, among others, depression, phobia and anxiety. The Conduct problems subscale focuses on behavioural disorders, the Hyperactivity/inattention subscale can be associated with the diagnosis of ADHD, while the Peer relationships problem and Prosocial behaviour subscales can be associated with all diagnoses [143].

Research Questions and Hypotheses

In this research, we formed the following research questions:

1. How can the therapeutic adherence of children with T1DM be described?
2. Are there any connections between the different mental health indexes and adherence?
3. Which mental health indexes have a significant impact on diabetes adherence?

Based on these research questions, we formulated the following hypotheses:

1. Concerning adherence, we hypothesise that significant differences can be detected by gender, age, family structure and parents' educational level according to which girls, older children, those living in an intact family and whose parents have a higher educational level can be characterised with higher adherence.
2. We hypothesise that well-being, self-rated health, psychological mood, satisfaction with life and health-related quality of life are in positive connection with adherence, while depression, somatic symptoms and various behavioural problems show a negative correlation with adherence.
3. We hypothesise that self-rated health, psychological well-being, satisfaction with life and health-rated quality of life has a significantly positive impact on adherence while depression, somatic symptoms and behavioural problems have a significantly negative effect on adherence.

The data were collected in an Excel database and were analysed with SPSS 22.0 for Windows statistical program. Concerning the results of the Kolmogorov-Smirnov test, the distribution of the data is not normal, which allows us to use non-parametric tests. For this reason, between-group comparisons were carried out by Mann-Whitney (in case of two groups), and Kruskal-Wallis tests (in case of three or more groups) and Spearman rank correlation was performed to detect the relationship between the examined variables. In order to measure the direction of the effect, linear regression analysis was applied.

RESULTS

The General Description of Adherence

One of the main aims of our research was to map the two sides of the phenomenon of therapeutic collaboration among children and adolescents with T1DM. This was carried out by examining the children's mental health indicators based on subjective self-reports, as well as their HbA1C values, which provide objective information about the average blood sugar level over the past three months.

First, we detected the general points of the subscales in the sample. The results can be seen in Fig. (**1**).

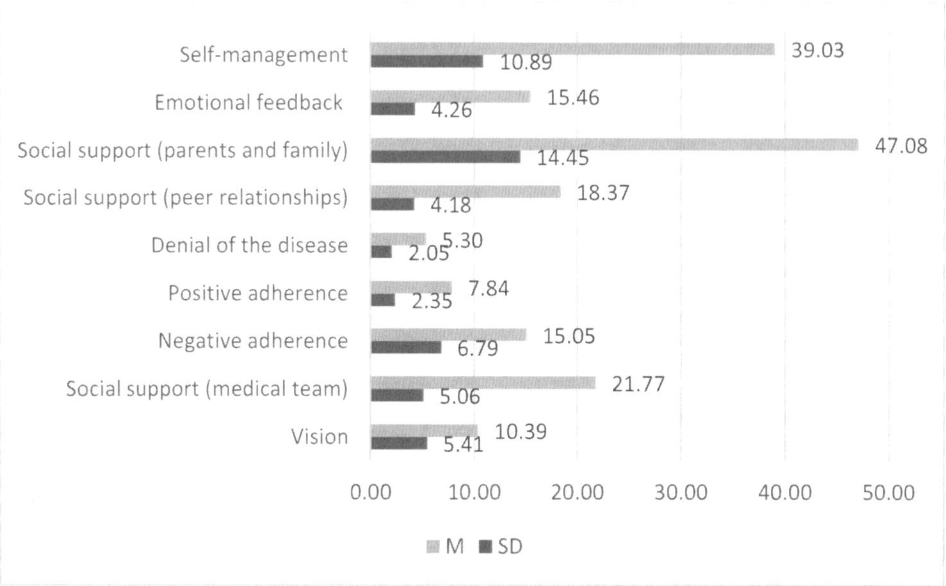

Fig. (1). The means of the Diabetes Adherence Questionnaire (DAQ, N=114).

In the case of T1DM, it is also important to examine the effect of sociodemographic variables. We first analysed the relationships between adherence (total score and subscale scores) and sociodemographic variables.

Examining the gender differences (Table **5**) with the Mann-Whitney test, we found that boys reached significantly higher points than girls in the total adherence score. In this case, significant differences could have been found concerning the Emotional feedback (EF) and Vision (v) subscales.

Table 5. Differences in the subscales of Diabetes Adherence Questionnaire by age (Source: DAS, N=11).

Subscales	9-12 years (N=34)		13-14 years (N=26)		15-16 years (N=32)		17-20 years (N=21)		p
	M	SD	M	SD	M	SD	M	SD	
Self-management	38,7	14,2	39,7	7,1	41,1	9,1	35,2	11,0	,049
Emotional feedback	15,0	5,8	15,9	1,9	15,4	3,9	15,5	4,1	,689
Social support (parents and family)	49,1	17,8	48,6	8,0	46,4	14,3	42,0	14,5	,014
Social support (peer relationships)	17,9	4,9	18,7	2,9	18,4	4,2	18,6	4,5	,943
Denial of the disease	5,1	2,2	5,3	1,9	5,6	2,2	5,3	1,8	,797

(Table 5) cont.....

Subscales	9-12 years (N=34)		13-14 years (N=26)		15-16 years (N=32)		17-20 years (N=21)		p
	M	SD	M	SD	M	SD	M	SD	
Positive adherence	7,6	3,1	8,4	1,5	7,9	2,1	7,3	2,1	,215
Negative adherence	15,0	8,7	15,8	6,0	14,5	5,8	15,2	6,0	,767
Social support (medical team)	21,2	6,2	22,1	2,9	22,5	5,0	21,0	5,4	,191
Vision	9,5	7,0	10,2	3,7	10,8	5,0	11,6	5,0	,053
Total	179,1	57,2	184,8	14,7	182,7	37,0	171,8	42,5	,009

The sample can be divided into four well-distinguishable groups based on age, namely 9-12 years old (N = 34), 13-14 years old (N = 26), 15-16 years old (N = 32), and 17-20 years old (N = 21) children. We examined the differences in adherence by age, the results of which are illustrated in Table **6**. Based on the results, 13-14-year-olds have the highest level of adherence, while 17-20-ye-r-olds have the lowest level. However, the difference between groups is not significant for either total adherence or its subscales.

Table 6. Differences in the subscales of Diabetes Adherence Questionnaire by family structure (Source: DAS, N=11).

Family Structure	Intact (N=77)		Non-intact (N=25)		Foster Parents (N=10)		-
-	M	SD	M	SD	M	SD	p
Self-management	40,1	7,1	39,3	12,1	33,7	8,7	,029
Emotional feedback	16,2	2,5	15,5	4,6	13,5	5,5	,242
Social support (parents and family)	47,0	13,1	48,0	14,9	38,9	13,9	,062
Social support (peer relationships)	19,1	3,1	17,9	4,7	19,6	1,6	,374
Denial of the disease	5,2	1,8	5,2	2,2	6,3	1,4	,229
Positive adherence	8,1	1,5	7,9	2,5	6,3	2,5	,020
Negative adherence	16,8	6,8	14,5	6,8	15,3	6,6	,294
Social support (medical team)	22,4	3,1	21,6	5,7	21,0	4,3	,590
Vision	11,4	5,6	10,0	5,4	10,1	5,4	,664
Total	186,4	21,8	180,0	47,1	164,7	35,0	,021

Based on the family structure (Table **7**), it can be said that the highest adherence was characteristic of young people living in intact families, while the lowest adherence was found among children living with foster parents. However, the difference between the groups is significant only for the total adherence and social support and vision subscales.

Table 7. Differences in the subscales of Diabetes Adherence Questionnaire by the parents educational level (Source: DAS, N=11).

-	Father's educational level							Mother's educational level						
-	Primary level		Secondary level		Tertiary level		p	Primary level		Secondary level		Tertiary level		p
-	M	SD	M	SD	M	SD		M	SD	M	SD	M	SD	
Self-management	41,5	8,0	41,0	5,9	41,5	8,00	0,389	40,2	7,6	41,9	6,2	40,2	7,6	0,110
Emotional feedback	16,0	2,0	16,5	2,2	16,0	2,00	0,102	16,0	2,0	16,3	2,0	16,0	2,0	0,280
Social support (parents and family)	50,4	7,9	50,2	8,8	50,4	7,89	0,208	49,2	8,2	50,1	9,1	49,2	8,2	0,436
Social support (peer relationships)	19,5	2,0	19,2	1,9	19,5	1,97	0,147	19,3	1,9	19,0	2,2	19,3	1,9	0,448
Denial of the disease	4,9	1,9	6,0	1,9	4,9	1,87	0,180	5,3	2,1	5,8	1,9	5,3	2,1	0,273
Positive adherence	7,9	1,7	8,5	1,6	7,9	1,70	0,092	8,2	1,6	8,5	1,5	8,2	1,6	0,121
Negative adherence	15,7	8,3	14,2	4,9	15,7	8,33	0,339	16,0	8,4	14,9	5,5	16,0	8,4	0,836
Social support (medical team)	22,5	2,9	22,8	2,8	22,5	2,94	0,245	22,8	3,1	22,6	2,8	22,8	3,1	0,501
Vision	12,2	5,5	9,3	3,4	12,2	5,51	0,204	11,8	5,3	10,1	4,2	11,8	5,3	0,344
Total	190,5	23,4	187,8	14,2	190,5	23,39	0,296	188,7	23,0	189,2	14,3	188,7	23,0	0,206

Concerning the parents' educational level, no significant differences could have been detected between the mothers and the fathers (Table **8**). Generally, the same could have been seen concerning the presence of siblings as no significant difference could have been experienced in the overall adherence by having a sibling. However, the Emotional feedback subscale highlighted a significant difference, meaning that those with at least one biological or foster sibling can be characterised with better emotional feedback. Thus, our hypothesis was partly confirmed.

Table 8. Differences in the subscales of Diabetes Adherence Questionnaire by having at least one sibling (Source: DAS, N=11).

Subscales	No		Yes		p
	M	SD	M	SD	
Self-management	42,5	6,7	39,1	10,6	0,179
Emotional feedback	15,1	1,8	17,3	4,3	0,041
Social support (parents and family)	52,0	8,7	47,0	14,4	0,150
Social support (peer relationships)	18,7	3,2	18,4	4,3	0,797
Denial of the disease	5,2	1,8	5,4	2,1	0,792
Positive adherence	8,4	1,6	7,8	2,3	0,303
Negative adherence	16,4	8,0	14,9	6,5	0,386
Social support (medical team)	23,0	3,2	21,6	5,4	0,283
Vision	11,7	7,0	10,2	5,0	0,286
Total	194,9	13,5	179,7	42,5	0,127

Table 9. Relationships between adherence and the variables of mental health.

Correlation (r) and significance (p)	1. SM	2. EF	3. SSF	4. SSP	5. SSM	6. DD	7. PA	8. NA	9. V	Total
CDI	-0,1808 0,0779	-0,2469 0,0153*	-0,1397 0,1747	-0,3370 0,0008**	-0,2642 0,0093*	-0,1415 0,1690	0,390 p<0,001**	0,3261 0,0012*	-0,3821 0,0001**	-0,4019 p<0,001**
SWLS	0,3888 p<0,001**	0,2268 0,0183*	0,4054 p<0,001**	0,2639 0,0058*	0,3068 0,0012**	0,2101 0,0291*	-0,489 p<0,001**	0,4358 p<0,001**	0,5151 p<0,001**	0,564 p<0,001**
SWL-p	0,3963 p<0,001**	0,2002 0,0387*	0,2872 0,0027*	0,3046 0,0014	0,1851 0,0563	0,2549 0,0081	0,2899 0,0025	0,4055 p<0,001**	0.5465 p<0,001**	0.5153 p<0,001**
SWL-f	0,1959 0,0431	0,2087 0,0310	0,2685 0,0052	0,3181 0,0008	0,0686 0,4827	0,0545 0,5773	0,1799 0,0637	0,2771 0,0039**	0.3293 0,0005**	0,3235 0,0007**
WBI	0,3277; p<0,001**	0,1055; 0,2771	0,3423; 0,0003**	0,2360; 0,0166*	0,2004; 0,0375*	0,0913; 0,3471	-0,293 0,002**	0,3422 0,0003**	0,4720 p<0,001**	0,475 p<0,001**
Somatic symptoms	-0,2119; 0,0277*	-0,2708; 0,0046*	-0,1887; 0,0505*	-0,1852; 0,055*	-0,1502; 0,1207	-0,0642; 0,5092	0,073 0,618	-0,3165 0,0008**	-0,5107 p<0,001**	0,427 p<0,001**
PedsQL – self-rated	0,2343; 0,0147*	0,406; p<0,001**	0,2635; 0,0059*	0,3351; 0,0004**	0,2279; 0,0177*	0,1197; 0,2172	0,178 0,058	0,5629 p<0,001**	0,5046 p<0,001**	0,5528 p<0,001**
PedsQL - external assessment	-0,0146; 0,9152	0,2299; 0,0882	0,0537; 0,6945	0,2750; 0,0402*	-0,0039; 0,9774	0,0902; 0,5087	0,124 0,348	0,2307 0,0871	0,1424 0,2951	0,1983 0,1430
SDQ – self-rated	-0,3248; 0,0188*	-0,4082; 0,0027*	-0,3102; 0,0252*	-0,3597; 0,0088*	-0,2614; 0,0613	0,0390; 0,7839	0,333 0,112	-0,6262 p<0,001**	-0,4337 0,0013*	0,6536 p<0,001**
SDQ - extenrnal assessment	0,0522; 0,7336	-0,1759; 0,2478	-0,2247; 0,1377	-0,4007; 0,0064*	-0,0655; 0,6691	-0,0169; 0,9122	0,163 0,782	-0,1669 0,2732	-0,0925 0,5456	-0,279 0,0634

Spearman rank correlations, *p < 0,05; **p < 0,01.

The Relationship Between Adherence and the Indexes of Mental Health

Another main aim of our research was to map the connection between the indexes of mental health and adherence, carried out by Spearman rank correlation. Our results showed a significant correlation between adherence and depression, somatic complaints, life satisfaction, subjective well-being, self-esteem, quality of life, and emotional and behavioural symptoms (Tables **9** and **10**).

Table 10. The relationship between adherence and quality of life (PedsQL).

r, p	-	1. SM	2. EF	3. SSF	4. SSP	5. SSM	6. DD	7. PA	8. NA	9. V	Total
Child questionnaire	Physical Functioning	0,2455 0,0104*	0,3338 0,0004**	0,1877 0,0517*	0,2432 0,0112*	0,2568 0,0073*	0,0939 0,3339	0,110 0,335	0,5147 p<0,001**	0,4286 p<0,001**	0,4716 p<0,001**
	Emotional Functioning	0,1302 0,1793	0,4155 p<0,001**	0,2222 0,0208*	0,3356 0,0004**	0,1138 0,2409	0,1347 0,1645	0,008 0,942	0,5042 p<0,001**	0,5705 p<0,001**	0,4997 p<0,001**
	Social Functioning	0,2114 0,0281*	0,3866 p<0,001**	0,2727 0,0043*	0,4790 p<0,001**	0,2009 0,0371*	0,0797 0,4121	-0,104 0,360	0,5091 p<0,001**	0,3670 0,0001**	0,5114 p<0,001**
	School Functioning	0,2097 0,0294*	0,2567 0,0073*	0,2432 0,0112*	0,1248 0,1981	0,2299 0,0167*	0,1241 0,2006	-0,260 0,505	0,3720 0,0001**	0,3170 0,0008**	0,4009 p<0,001**
	Total PedsQL™ index	0,2343 0,0147*	0,0460 p<0,001**	0,2635 0,0059*	0,3351 0,0004**	0,2279 0,0177*	0,1197 0,2172	0,297 0,139	0,5629 p<0,001**	0,5046 p<0,001**	0,5528 p<0,001**
Parental questionnaire	Physical Functioning	-0,0604 0,6583	0,0717 0,5996	-0,1144 0,4012	0,0836 0,5400	-0,1317 0,3331	-0,0516 0,7058	0,038 0,744	-0,0165 0,9040	0,0210 0,8781	-0,045 0,7390
	Emotional Functioning	-0,0266 0,8457	0,3959 0,0025*	0,1766 0,1929	0,2957 0,0269*	-0,0067 0,9611	0,0890 0,5142	-0,110 0,337	0,2608 0,0522*	0,2960 0,0268*	0,3408 0,0102*
	Social Functioning	-0,1113 0,4142	0,1259 0,3553	-0,0483 0,7238	0,3358 0,0114*	-0,0602 0,6596	0,0517 0,7053	0,020 0,863	0,1097 0,4207	0,2127 0,1156	0,0898 0,5102
	School Functioning	0,0287 0,8337	0,1241 0,3622	0,0456 0,7385	0,0620 0,6498	-0,0039 0,9774	0,0180 0,8955	-0,105 0,358	0,1443 0,2885	0,0074 0,9566	0,0486 0,7219
	Total PedsQL™ index	-0,0146 0,9152	0,2299 0,0882	0,0537 0,6945	0,2750 0,0402*	-0,0039 0,9774	0,0902 0,5087	0,125 0,485	0,2307 0,0871	0,1424 0,2951	0,1983 0,1430

Spearman rank correlations, *p < 0,05; **p < 0,01

The correlations between adherence, age, the appearance of symptoms, time of the diagnosis and HbA1c level were investigated with Spearman rank correlation. The overall adherence did not show any significant correlation with the above mentioned diabetes-specific variables (age: r=-0.06, p=0.949; appearance of symptoms: r=-0.172, p=0.355; age at the diagnosis: r=-0.193, p=0.428; duration of T1DM: r=-0.108, p=0.253; HbA1c: r=-0.120, p=0.208). No significant correlation was found concerning the subscales of adherence. Our results showed a significant positive correlation for the Social support of peers subscale (r=0.203; p=0.019). Thus, children with a longer history of diabetes had a significantly higher result on this subscale. For the Negative adherence subscale, the results show a tendency-level connection with the age of the diagnosis (r=0.301;

p=0.062). Thus, those diagnosed at a younger age scored slightly higher on this subscale and are less affected by the emotional burden of diabetes care.

We found a significant correlation between the subscales of our adherence questionnaire (DAQ) and depression (CDI) concerning the Emotional feedback, Social Support - Peers, Social Support - Medical Team, Negative Adherence, and Vision subscales. Of these, the strongest correlations were found in the Social Support - Peers and Vision subscales. Thus, those children, who found their treatment tasks more emotionally burdensome, who felt that living with diabetes had a negative impact on their peer relationships, who felt that the medical team was less reliable and supportive, for whom treatment tasks were significantly more prohibitive, and whose concerns about the future were more significant reported considerably higher levels of depression.

There was a significant positive relationship for all subscales in Satisfaction with Life (SWLS, SWL-p). The most significant correlations were found concerning the Self-management, Social Support - Family, Negative Adherence, and Vision subscales. Mental well-being (WBI) showed a significant positive correlation with Self-management, all three dimensions of Social Support (Family, Peers, Medical Team), Negative Adherence, and Vision subscales. Significant correlations with the frequency of somatic complaints were found for the Self-management, Emotional feedback, two Dimensions of Social Support (Family and Peers), Negative Adherence, and Vision subscales. The most significant correlations were in the case of the Negative Adherence and Vision subscales.

The overall score of our adherence questionnaire showed a significant correlation with children's self-rated mental health status variables, while not with parental assessments. The strongest correlations were with emotional and behavioural symptoms (SDQ), satisfaction with life (SWLS), and quality of life (PedsQL) (Table **6**).

Self-rated quality of life (PedsQL) showed a significant correlation with all subscales except the Denial of the disease. The strongest correlations were found with the Emotional feedback, Social Support (Peers), Negative Adherence, and Vision subscales. There was a significant relationship between parents' quality of life and the Social Support (Peers) subscale, meaning that parents of children who considered their peer relationships less negatively affected by diabetes were significantly more positive about their parents' quality of life.

The overall SDQ problem score showed a significant positive correlation with the Self-Management, Emotional Feedback, Social Support (Family and Peer), Negative Adherence, and Vision subscales. This means that children who scored significantly lower on emotional and behavioural problems scored more positively on their therapeutic collaboration and family support, emotional strain and limitations in treatment tasks, less negative impact on diabetes on their peer relationships, and fewer concerns about the future. Of these, the strongest correlations were with the Emotional Feedback and Negative Adherence subscales.

We measured health-related quality of life in a more in-depth way (Table **11**). The total score of our adherence questionnaire showed a strong positive significant correlation for the overall PedsQLTM index of the children-assessed version and all four domains of quality of life (physical, emotional, social, and school functioning). We also examined the relationships between adherence and quality of life rated by the parents. Regarding parental evaluation, adherence only showed a significant positive correlation with the child's emotional status, which shows that the evaluations of the parents and the children are not always consistent as they have different perceptions on the above-mentioned areas. The child's assessed emotional state showed a significant positive correlation with the Emotional Feedback, Social Support (Peers), and Vision subscales. Social function showed a significant association with the Social Support (Peers) subscale.

Table 11. The relationship between adherence and Strengths & Difficulties (SDQ).

-	-	1. SM	2. EF	3. SSF	4. SSP	5. SSM	6. DD	7. PA	8. NA	9. V	Total
Child questionnaire	Emotional symptoms	0,0882 0,5341	-0,4687 0,0003**	-0,1607 0,2550	-0,4040 0,0030*	-0,1274 0,3679	0,0270 0,8494	0,038 0,744	-0,5740 p<0,001**	-0,4493 0,0008**	0,5401 p<0,001**
	Conduct problems	-0,2970 0,0325*	-0,1696 0,2293	-0,1388 0,3263	-0,0885 0,5327	0,1239 0,3815	-0,0361 0,7992	-0,110 0,337	-0,2765 0,0473*	-0,2704 0,0526*	0,3646 p<0,0079*
	Hyperactivity/inattention	-0,3880 0,0045*	-0,2073 0,1403	-0,2577 0,0651	-0,0825 0,5608	-0,3785 0,0057*	-0,0127 0,9285	0,020 0,863	-0,3682 0,0072*	-0,1006 0,4777	0,3451 p<0,0122*
	Peer relationships problem	-0,2188 0,1191	-0,1832 0,1936	-0,3950 0,0038*	-0,2434 0,0821	-0,1582 0,2628	0,2103 0,1345	-0,105 0,358	-0,3826 0,0051*	-0,3214 0,0202*	0,5269 p<0,001**
	Prosocial behaviour	0,1440 0,3086	-0,0643 0,6508	0,3780 0,0057*	0,0277 0,8455	0,0812 0,5672	-0,0761 0,5918	-0,041 0,722	-0,0201 0,8878	-0,0934 0,5102	0,1623 0,2502
	Total	-0,3248 0,0188*	-0,4082 0,0027*	-0,3102 0,0252*	-0,3597 0,0088*	-0,2614 0,0613	0,0390 0,7839	-0,105 0,358	-0,6262 p<0,001**	-0,4337 0,0013*	0,6536 p<0,001**
Parental questionnaire	Emotional symptoms	0,0787 0,6075	-0,3037 0,0425*	-0,1004 0,5118	-0,3828 0,0094*	-0,0167 0,9131	-0,0960 0,5304	0,140 0,453	-0,2720 0,0707	-0,2953 0,0489*	0,3033 0,0428*
	Conduct problems	-0,0709 0,6436	0,0023 0,9883	-0,1922 0,2059	-0,3219 0,0311*	-0,0253 0,8692	-0,1245 0,4152	0,245 0,177	-0,0994 0,5158	-0,1473 0,3342	0,2524 0,0944
	Hyperactivity/inattention	0,0024 0,9874	-0,0015 0,9921	-0,1251 0,4129	-0,0602 0,6945	-0,0184 0,9043	0,0246 0,8724	0,078 0,675	0,1313 0,3900	0,2784 0,0640	0,0074 0,9617
	Peer relationships problem	0,1202 0,4315	-0,1047 0,4936	-0,2209 0,1448	-0,3861 0,0088*	-0,0268 0,8614	0,1447 0,3429	0,132 0,479	-0,0570 0,7100	-0,0524 0,7325	0,2236 0,1399
	Prosocial behaviour	0,1187 0,4372	0,1483 0,3309	0,4818 0,0008**	0,3224 0,0308*	0,0504 0,7422	0,0992 0,5166	0,353 0,047*	-0,0540 0,7246	0,0184 0,9046	0,2987 0,0463*
	Total	0,0522 0,7336	0,1759 0,2478	-0,2247 0,1377	-0,4007 0,0064*	-0,0655 0,6691	-0,0169 0,9122	0,402 0,023*	-0,1669 0,2732	-0,0925 0,5456	0,279 0,0634

Spearman- rank correlations, *p < 0,05; **p < 0,01.

Physical functioning and Social functioning showed a positive relationship with Self-management, Emotional feedback, Social support (family, peers and medical team) and negative adherence. Emotional functioning significantly correlated with Emotional feedback, Social support (family and peers), negative adherence and visions. School functioning showed a positive connection with Self-management, Emotional Feedback, Social support (parents and medical team), Negative adherence and Visions.

Concerning the assessment of the parents, different results could have been obtained. Physical functioning and school functioning did not show any correlation with the subscales of adherence. Emotional functioning correlated with Emotional feedback, Social support (peers), negative adherence and visions. Social functioning showed a positive relationship with Social support (peers).

We also examined the associations between adherence and self-assessed emotional and behavioural symptoms (SDQ) (Table **12**). We found a significant negative correlation between children's emotional symptoms and Diabetes-specific emotional feedback, Social Support (peers), Negative Adherence, and Vision subscales. There was a significant negative correlation between behavioural problems and the Self-management, Negative adherence, and Vision subscales. Hyperactivity showed a significant negative correlation with the Self-management, Social Support (medical team), and Negative Adherence subscales. Regarding Peer Relationship Problems, there was a significant negative correlation between the Social support (Family), Negative adherence, and Vision subscales. The overall adherence index showed a significant negative correlation with all subscales except Social Support (medical team) and Denial of the disease. Of these, the strongest relationships were found with the Negative Adherence and Vision subscales. Prosocial behaviour showed a significant positive correlation with the Social Support (Family) subscale.

Table 12. Factors influencing diabetes adherence based on the results of the linear regression analysis (expβ, Source: DAF 2017; N=114).

	Model1	Model2	Model3
Self-rated health (SRH)	,412***	,364**	,341**
Headache	-,782***	-,698***	-,657***
Diarrhea	-,497***	-,396***	-,378***
Backache	-,021	,011	-,120
Listlessness	-,787***	-,874***	-,854***
Sleeping difficulties	-,669***	-,663***	-,623***
Dizzyness	-,372**	-,351**	-0,357**

(Table 12) cont.....

	Model1	Model2	Model3
Tiredness	-,348***	-,262***	-,264**
Psychological wellbeing (WBI)	-	,159***	,184***
Depression (CDI)	-	-,337**	-0,162*
Satisfaction with life (SWLS)	-	-,052	-0,052
Quality of Life (PedSQL)	-	-	,273***
Stengtheness and difficulties (SDQ)	-	-	-0,10
SWL-p	-	-	0,204
SWL-f	-	-	0,127
Body image	-	-	0,105

***p<0,001

There were significant negative correlations between the total score of our adherence questionnaire and both the overall score and the individual sub-domains of the self-assessed version of the SDQ for children. Of these, the most significant correlations existed between the overall problem score and the emotional and psychosocial domains. Examining the correlations between adherence and emotional and behavioural symptoms perceived by parents, we found a significant negative correlation between emotional symptoms and Emotional feedback, Social Support (peers) and Vision subscales. Regarding behavioural problems, the Social Support (peers) subscale showed a significant negative correlation. No significant association could have been detected between Hyperactivity and the subscales of adherence. We found a significant negative correlation with the Social Support (Family and Peers) subscales regarding prosocial behaviour. Thus, as it could have been claimed in the case of the health-related quality of life, the evaluation of the parents are not in a strong correlation with the diabetes adherence of the children. Parents often have different perceptions of their child compared to how the child evaluates things.

Based on our results, lower levels of adherence were indeed correlated with more negative mental health indicators. Thus, our hypothesis was partly confirmed. Mental health is a complex system, so a problem in one of its components or a negative change results in a negative shift in the case of other mental health factors as well.

Psychological Factors Influencing Adherence

Finally, we examined the effect of the psychological factors on adherence, the results of which are illustrated in Table **13**. We involved the psychological factors through three models (Table **14**).

The English Version of the Diabetes Adherence Questionnaire

Table 13. You can read statements related to Type 1 Diabetes and its treatment. Please, describe how typical these statements are for you (1 = not typical at all, 5 = absolutely typical).

-	-	Not Typical at all	-	-	-	Absolutely Typical
1	My treatment causes inconvenience.	1	2	3	4	5
2	My treatment requires too much time and work.	1	2	3	4	5
3	Nothing wrong can happen to me if I do not follow the treatment schedule.	1	2	3	4	5
4	The relationship with friends is more important than treating diabetes.	1	2	3	4	5
5	I try to forget diabetes.	1	2	3	4	5
6	I am discriminated in my friendships due to diabetes.	1	2	3	4	5
7	I trust in doctors.	1	2	3	4	5
8	Doctors are too busy to discuss diabetes with me.	1	2	3	4	5
9	Doctors understand that diabetes hinders me from doing things that are important for me.	1	2	3	4	5
10	Doctors are friendly and can easily talk to me.	1	2	3	4	5
11	I trust in nurses.	1	2	3	4	5
12	I follow the suggestions of the schedule.	1	2	3	4	5
13	My family encourages me to keep my diet.	1	2	3	4	5
14	My family remembers me to keep my diet.	1	2	3	4	5
15	My family members control my appropriate consumption of meals	1	2	3	4	5
16	My family members prepare dietary meals for me.	1	2	3	4	5
17	My family members commend me when I keep my diet and eat healthily.	1	2	3	4	5
18	My family remembers me when I eat inappropriate food.	1	2	3	4	5
19	I do enough physical activity.	1	2	3	4	5
20	My family commends me for doing physical activity.	1	2	3	4	5
21	My family members encourage me to pursue sport.	1	2	3	4	5
22	I avoid my peers becoming aware of my diabetes if possible.	1	2	3	4	5
23	It is disturbing when my friends ask about diabetes.	1	2	3	4	5
24	Most of my friends are not diagnosed with diabetes; thus, I feel different.	1	2	3	4	5
25	I prefer to let my family members communicate with my environment about diabetes.	1	2	3	4	5

(Table 13) cont.....

-	-	Not Typical at all	-	-	-	Absolutely Typical
26	I prefer to communicate with my environment about diabetes on my own.	1	2	3	4	5
27	I eat an appropriate amount of food.	1	2	3	4	5
28	I eat at an appropriate time (breakfast, snack, lunch *etc.*).	1	2	3	4	5
29	I take an appropriate amount of insulin.	1	2	3	4	5
30	I take insulin in the appropriate time.	1	2	3	4	5
31	When taking insulin, I take my blood pressure into account.	1	2	3	4	5
32	It evokes guilt when my blood sugar is high.	1	2	3	4	5
33	It evokes fear when my blood sugar is low.	1	2	3	4	5
34	I measure my blood glucose level in all circumstances.	1	2	3	4	5
35	I record my blood glucose level regularly.	1	2	3	4	5
36	I do my best to control and keep diabetes in line.	1	2	3	4	5
37	Diabetes limits my friendships.	1	2	3	4	5
38	Diabetes strongly limits my lifestyle.	1	2	3	4	5
39	The appropriate treatment of diabetes requires a huge amount of time.	1	2	3	4	5
40	My family is glad when I appropriately treat diabetes.	1	2	3	4	5
41	The more I follow my treatment schedule, the better I feel.	1	2	3	4	5
42	My diet significantly limits my lifestyle.	1	2	3	4	5
43	Diabetes constrains me from pursuing sports.	1	2	3	4	5
44	Diabetes limits my leisure activities.	1	2	3	4	5
45	Diabetes limits my school activities.	1	2	3	4	5
46	I am afraid whether I will marry due to diabetes.	1	2	3	4	5
47	I am afraid whether I will have any child due to diabetes.	1	2	3	4	5
48	I am afraid that it will be harder to find a job due to diabetes.	1	2	3	4	5
49	I am afraid that I cannot finish my studies due to diabetes.	1	2	3	4	5
50	I am afraid that diabetes influences my physical appearance.	1	2	3	4	5
51	I am afraid that I will have medical complications due to diabetes.	1	2	3	4	5
52	I can control my fears related to diabetes.	1	2	3	4	5
53	During meals, I eat an appropriate amount of food following my diet.	1	2	3	4	5
54	My family commends me when following the schedule.	1	2	3	4	5
55	My family members help me in changing the treatment on the basis of my blood glucose.	1	2	3	4	5

(Table 13) cont.....

-	-	Not Typical at all	-	-	-	Absolutely Typical
56	I become afraid because of my high level of blood glucose.	1	2	3	4	5
57	I become afraid because of my low level of blood glucose	1	2	3	4	5
58	Sometimes I forget the treatment of diabetes.	1	2	3	4	5

Evaluation

1. Self-management: 12, 27, 28, 29, 30, 31, 34, 35, 36, 53

2. Emotional feedback (emotional reactions associated with blood sugar level measurement): 25, 26, 32, 33, 52, 56, 57

3. Social support (parents and family): 13, 14, 15, 16, 17, 18, 20, 21, 40, 54, 55

4. Social support (peer relationships): 4, 6, 22, 23, 24,

5. Denial of the disease: 3, 5, 58

6. Positive adherence: 19, 41

7. Negative adherence: 1, 2, 37, 38, 39, 42, 43, 44, 45,

8. Social support (medical team): 7, 8, 9, 10, 11

9. Vision: 46, 47, 48, 49, 50, 51

The Original Version of the Diabetes Adherence Questionnaire

Table 14. A továbbiakban a cukorbetegséggel és annak kezelésével kapcsolatos állításokat olvashatsz. Kérlek, állapítsd meg, mennyire jellemzőek rád ezek az állítások! Az 1-es az egyáltalán nem, míg az 5-ös a teljes mértékben jellemző.

-	-	Egyáltalán nem	-	-	-	Teljes Mértékben
1	Kezelésem kellemetlenséget okoz.	1	2	3	4	5
2	Kezelésem sok időt és munkát igényel.	1	2	3	4	5
3	Semmi rossz nem történhet velem, ha nem tartom be a kezelési rendet.	1	2	3	4	5
4	A barátokkal való kapcsolat fontosabb, mint cukorbetegségem kezelése.	1	2	3	4	5
5	Próbálom elfelejteni cukorbetegségemet.	1	2	3	4	5

(Table 14) cont.....

-	-	Egyáltalán nem	-	-	-	Teljes Mértékben
6	Cukorbetegségem miatt megkülönböztetnek baráti társaságomban.	1	2	3	4	5
7	Megbízom az orvosokban.	1	2	3	4	5
8	Az orvosok túlságosan elfoglaltak ahhoz, hogy beszéljenek velem cukorbetegségemről.	1	2	3	4	5
9	Az orvosok megértik, hogy cukorbetegségem mennyire akadályoz a számomra fontos dolgokban.	1	2	3	4	5
10	Az orvosok barátságosak és könnyen tudnak velem beszélgetni.	1	2	3	4	5
11	Megbízom a nővérekben.	1	2	3	4	5
12	Betartom a kezelési előírásokat.	1	2	3	4	5
13	Családom bíztat diétám betartására.	1	2	3	4	5
14	Családom emlékeztet diétám betartására.	1	2	3	4	5
15	Családtagjaim kontrollálják megfelelő ételfogyasztásomat.	1	2	3	4	5
16	Családtagjaim készítik el a diétás ételeket számomra.	1	2	3	4	5
17	Családtagjaim megdicsérnek, ha megfelelően étkezem, betartom a diétámat.	1	2	3	4	5
18	Családom figyelmeztet, ha nem megfelelő ételt eszem.	1	2	3	4	5
19	Eleget mozgok, sportolok.	1	2	3	4	5
20	Családom megdicsér a fizikai aktivitásért, sportért.	1	2	3	4	5
21	Családtagjaim sporttevékenységek végzésére bátorítanak.	1	2	3	4	5
22	Ha lehet, elkerülöm, hogy társaim tudjanak cukorbetegségemről.	1	2	3	4	5
23	Zavar, ha barátaim cukorbetegségemről kérdeznek.	1	2	3	4	5
24	Barátaim többsége nem cukorbeteg, ezért másnak érzem magam.	1	2	3	4	5
25	Jobban szeretem, ha családtagjaim közlik helyettem környezetemmel cukorbetegségemet.	1	2	3	4	5
26	Jobban szeretem, ha én közlöm környezetemmel cukorbetegségemet.	1	2	3	4	5
27	Megfelelő ételmennyiséget eszek.	1	2	3	4	5
28	Étkezéseimet (reggeli/tízórai/ebéd stb.) megfelelő időben végzem.	1	2	3	4	5
29	Megfelelő inzulinmennyiséget adok be magamnak.	1	2	3	4	5
30	Megfelelő időben adom be magamnak az inzulint.	1	2	3	4	5
31	Vércukormérésem eredményét figyelembe véve adom be magamnak az inzulint.	1	2	3	4	5
32	Bűntudatot vált ki, ha magas a vércukorszintem.	1	2	3	4	5
33	Félelmet vált ki, ha alacsony a vércukorszintem.	1	2	3	4	5
34	Minden körülmények között megmérem a vércukromat.	1	2	3	4	5

(Table 14) cont.....

-	-	Egyáltalán nem	-	-	-	Teljes Mértékben
35	Rendszeresen rögzítem vércukorszintem mérésének eredményeit.	1	2	3	4	5
36	Mindent megteszek, hogy kordában tartsam cukorbetegségemet.	1	2	3	4	5
37	Cukorbetegségem korlátozza baráti kapcsolataimat.	1	2	3	4	5
38	Cukorbetegségem nagyban korlátozza életvitelemet.	1	2	3	4	5
39	Cukorbetegségem megfelelő kezelése sok időt vesz igénybe.	1	2	3	4	5
40	Családom örül, ha megfelelően kezelem magam.	1	2	3	4	5
41	Minél jobban betartom kezelési előírásaimat, annál jobban érzem magam.	1	2	3	4	5
42	Diétám jelentősen korlátoz életvezetésemben.	1	2	3	4	5
43	Cukorbetegségem korlátoz a mozgásban, sportolásban.	1	2	3	4	5
44	Cukorbetegségem korlátozza szabadidős tevékenységeimet.	1	2	3	4	5
45	Cukorbetegségem korlátozza iskolai tevékenységeimet.	1	2	3	4	5
46	Aggódom, hogy cukorbetegségem miatt megházasodom-e.	1	2	3	4	5
47	Aggódom, hogy lehet-e majd gyermekem a cukorbetegségem miatt.	1	2	3	4	5
48	Aggódom, hogy nehezebb lesz majd állást találnom cukorbetegségem miatt.	1	2	3	4	5
49	Aggódom, hogy nem tudom majd befejezni iskolai tanulmányaimat cukorbetegségem miatt.	1	2	3	4	5
50	Aggódom, hogy cukorbetegségem testi megjelenésemet is befolyásolja.	1	2	3	4	5
51	Aggódom, hogy egészségügyi komplikációim adódnak cukorbetegségem miatt.	1	2	3	4	5
52	Cukorbetegségemmel kapcsolatos félelmeimet kordában tudom tartani.	1	2	3	4	5
53	Egy étkezés alkalmával diétámnak megfelelő ételmennyiséget eszek meg.	1	2	3	4	5
54	Családom megdicsér, ha betartom a kezelési előírásokat.	1	2	3	4	5
55	Családtagjaim segítenek a kezelés változtatásában vércukor mérésem alapján.	1	2	3	4	5
56	Magas vércukorszintem miatt megijedek.	1	2	3	4	5
57	Alacsony vércukorszintem miatt megijedek.	1	2	3	4	5
58	Időnként megfeledkezem cukorbetegségem kezeléséről.	1	2	3	4	5

Értékelési útmutató:

1. Önmenedzselés (self-management): 12, 27, 28, 29, 30, 31, 34, 35, 36, 53
2. Emocionális feedback (a vércukorméréssel kapcsolatos érzelmi reakciók): 25, 26, 32, 33, 52, 56
3. Társas támasz (szülő és család): 13, 14, 15, 16, 17, 18, 20, 21, 40, 54, 55
4. Társas támasz (kortársak): 4, 6, 22, 23, 24,

5. Betegségtagadás: 3, 5, 58
6. Pozitív adherencia: 19, 41
7. Negatív adherencia 1, 2, 37, 38, 39, 42, 43, 44, 45,
8. Társas támasz (orvosi team): 7, 8, 9, 10, 11
9. Jövőkép: 46, 47, 48, 49, 50, 51

In the first model, we examined the effect of *self-rated health status* and *psychological and somatic symptoms* on adherence. Based on the results of the linear regression, the effect of self-rated health status alone ($p=0.001$) and with the inclusion of psychological variables is significant ($p=0.01$ and $p=0.022$) and positive. Accordingly, better-rated health has a positive effect on adherence. In contrast, the effect of frequent headaches is negative and significant ($p<0.001$), suggesting that more frequent headaches have a negative effect on therapeutic adherence. A similar picture can be seen in the case of diarrhea, as its effect is significant ($p<0.001$) and negative. Significant negative effects were also observed in the case of listlessness ($p<0.001$), sleeping difficulties ($p<0.001$), dizziness ($p=0.035$), and tiredness ($p<0.001$). Negative bodily symptoms and negatively judged health status thus have a negative impact on therapeutic adherence.

In the second model, we examined the effect of *well-being, depression,* and *satisfaction with life* on adherence. The effect of well-being on adherence is significant ($p<0.001$) and positive, so a higher level of well-being results in a higher level of adherence. The effect of depression is also significant ($p<0.001$) but negative, so a higher level of depression weakens therapeutic adherence. The effect of satisfaction with life was not significant ($p=0.674$). The effect of self-rated health status and psychological and somatic symptoms examined in the previous model remains significant (except for back pain).

Finally, in our third model, we included and examined *health-related quality of life, strengthens and difficulties, the Cantril ladder* in the present and future terms, and *body image variables* in terms of adherence. Of the variables included in the model, only the effect of health-related quality of life was significant ($p<0.001$) and positive, suggesting that higher health-related quality of life leads to a higher level of therapeutic adherence. In the other cases, no significant effect was detected (strength and difficulties: $p=0.697$; Cantril ladder present time: $p=0.384$; Cantril ladder future time: $p=0.297$; body image: $p=0.361$). However, the impact of self-rated health status and psychological and somatic symptoms remained significant, and the same was true for depression and well-being.

To sum up, it can be stated that a better self-rated health condition has a significantly positive effect on adherence, and a similarly positive effect can be found in the case of well-being. Thus, positive psychological factors such as self-rated health and well-being appear to be able to move adherence in a positive direction. However, the effect of negative physical symptoms, psychosomatic symptoms and a negative health condition is significantly negative on therapeutic adherence, and the same can be seen in the case of depression. Thus, these variables deteriorate the level of adherence. Based on these results, neither negative nor positive effects can be assumed in terms of satisfaction with life, strengthens and difficulties, present and future visions, and body image. Thus, our hypothesis was partly confirmed.

SUMMARY

Given the effects of diabetes as a chronic and incurable disease on physical and mental health, the presence of more negative mental health indicators may seem evident when compared to healthy peers. However, due to the strict rules and changed and healthier lifestyle, well-being, mental health, somatic complaints, depression, and quality of life can be even better among children and adolescents with T1DM. As a result of the illness, children can learn to detect consciously the causes and possible treatments of the mental disparities, which can make them more effective and can lead to a more positive picture in terms of health-related quality of life, as well as self-image and self-esteem. As can be seen from further results, stable, resilient internal resource-based coping can be assumed in the background, as children diagnosed with T1DM have to learn to live with a chronic and incurable disease. If it is successful, based on post-traumatic growth, it results in additional resources and the acquisition of more effective coping strategies, which can have a positive effect on other mental health factors as well [144].

In studies using the Child Depression Questionnaire, nearly a quarter of the T1DM children studied (22.8% [145], 23% [146], 25% [147]) were at risk for depression or showed clinical-level depressive symptoms emphasising the relevance of this research topic. De Wit and Snoek [148] found that the rate of clinical-level depressive symptoms was 17.2% among T1DM children.

Self-rated health is another crucial topic to be investigated. According to the results of the Health Behaviour in School-aged Children (HBSC, 2014) research, 28.9% of Hungarian children and adolescents are excellent, 51.4% are good, but 17% are only suitable, 2.8% - classified his health as poor. Based on our previous results, it can be seen that the self-rated health status of T1DM children is significantly worse than the results of the nationally-representative sample used in the HBSC research, so they are a key risk group for lower-level health indicators. Along with the gender difference described in the HBSC research, more boys considered their health to be excellent, while more girls marked the other three response options. However, in the present study, there was no gender difference among T1DM children. Studies among adults with previous large sample numbers as protective factors for self-rated health [149, 150] have identified female gender, shorter duration of T1DM, younger age at occurrence of the disease, lower HbA1C levels, and better self-rated health status. However, in the present study, we did not find similar correlations, which can probably be explained by the small sample size and the difference in the age group.

Comparing the satisfaction with life of T1DM children with the Hungarian results of HBSC (2014), we can say that the life satisfaction of children with diabetes is almost the same. From our results, we can conclude that T1DM children are as satisfied with their lives as children representing the nationally representative sample, *i.e.*, they experience T1DM not as a limiting factor limiting their living space but as a state that can be well controlled by appropriate self-management.

Sociodemographic Factors Influencing Adherence

Examining the relationship between therapeutic adherence and glycemic control with sociodemographic factors showed that children's overall adherence scores, subscale scores and HbA1C values were not significantly related to sociodemographic factors (age, siblings, family structure, educational level of parents). However, concerning mental health factors, it is important to note that girls are usually characterised by significantly higher levels of depression and reported more frequent somatic complaints and poorer quality of life than boys. The increased incidence of depression in T1DM adolescent girls has been demonstrated in several studies [151]. Comorbid depressive symptoms in adolescent girls may contribute significantly to a decrease in therapeutic cooperation. This result is particularly important given the intensive behavioural therapeutic cooperation required by diabetes management and the potential for severe complications due to inadequate glycemic control. Also, adolescence is the developmental psychological period in which the developed therapeutic cooperation patterns are likely to perform in adulthood. Thus, depression can be identified as a risk factor for inadequate diabetes self-management, especially in adolescent girls. This effect is observed on self-reported adherence but not on objective metabolic control. Noteworthy is the result of La Greca *et al.* [127] that depression is a mediating factor for gender differences in metabolic control but does not appear in the case of adherence. Furthermore, glycemic control may be affected by several other factors and adherence and may even be particularly unstable in adolescence due to hormonal transformations. This lability may mask the effect of depression on metabolic control [152].

Our results that gender and age were not associated with either therapeutic collaboration or metabolic control are consistent with the research of Korbel *et al.* [152]. However, while they did not find a gender difference in depression, our results showed that girls had significantly higher levels of depression than boys, as did the HBSC research [153].

Hassan *et al.* [154] found a significant association between inadequate glycemic control and depression is more common among those with lower socioeconomic status. Inadequate glycemic control was associated with lower socioeconomic status and higher levels of depression. As glycemic control decreased, the likelihood of developing depression increased.

Regarding gender differences in therapeutic collaboration, our results showed that boys scored significantly higher on the total adherence scale, emotional feedback, and visions subscales but did not have significantly better metabolic control than girls. Thus, according to our interpretation, boys were characterised by significantly better therapeutic cooperation. For them, tasks and events related to diabetes were less emotionally burdening, and they reported fewer worries about the future than girls. Also, their glycemic control was closer to the reference zone provided by ISPAD. Our findings are consistent with international results. Adolescent girls typically show poorer metabolic control and are characterised by less appropriate therapeutic interactions [155] than boys. Gender differences of this nature are likely to reflect the diversity of changes during adolescence [156].

Based on our results, it can be stated that the effect of the demographic background variables included in the study, *i.e.* gender, family structure, siblings and parents' educational level is not significant, and the difference between the groups is not significant either. In the case of siblings, we can see that although the adherence of those having at least one sibling is higher compared to only children, the effect of siblings on adherence is not significant. Regarding the role of age, although its effect on adherence is not significant, children under 14 years of age have significantly lower adherence compared to older ones, confirming previous findings in the literature that adolescents have more effective coping potential [157, 158].

Psychological Factors Influencing Adherence

Factors related to nonadherence have been identified by several studies and have been divided into three groups: (1) characteristics of medical treatment, (2) illness characteristics, and (3) patient and/or family-related factors. The literature confirms a significant relationship between inadequate therapeutic collaboration and longer duration of the disease, complexity of treatment regimen, lifestyle changes, side effects, and uncertain efficacy in terms of the characteristics of medical treatment [159]. Inconsistent supervision by the medical team and parents may also contribute to inadequate therapeutic collaboration among adolescents [90]. According to the results of Rapoff and Barnard [160], a significant association was detected between nonadherent behaviour and younger age at the occurrence of the illness and family-perceived illness severity in terms of disease-specific factors. According to the results of Hood *et al.* [161], emotional reactions related to blood glucose measurement showed a significant negative relationship with quality of life. Following this, the more negative the child's emotional reaction, the worse the quality of life.

Contrary to expectations, we did not find a significant relationship between adherence and diabetes-specific variables. However, previously, Korbel *et al.* [152] found a significant negative association between the duration of T1DM and adherence, according to which the longer duration of diabetes was associated with poorer adherence and metabolic control. Kovacs *et al.* [162] further found that younger age at the diagnosis was a risk factor for inadequate adherence and medical and psychological problems. However, our results showed that the duration of T1DM was only significantly correlated with the Social support (peers) subscale, *i.e.*, children with longer-term T1DM were less aware of the limiting negative effects of diabetes on their peer relationships.

When examining the relationships between adherence and mental health status factors, we found significant relationships. Children reporting higher levels of therapeutic cooperation were characterised by lower levels of depression and emotional and behavioural symptoms, significantly fewer somatic complaints, but were more satisfied with their lives and reported higher levels of subjective well-being and quality of life. This result is consistent with previous research where it was found that depression correlated with significantly weaker adherence but showed only a trend-level positive relationship with metabolic control [152].

For children with lower levels of depression, living with diabetes is less emotionally stressful. They feel less negatively affected by their peer relationships, the medical team is evaluated more reliable, the burden of treatment is subjectively lower, and they have less concern about the future. T1DM children reporting higher levels of satisfaction with life rated their self-management significantly better, reported less diabetes-specific emotional strain and anxiety, and received significantly more positive support from their families, peer relationships, and the medical team, and they were less characterised by the Denial of the disease.

CONCLUSION

Our chapter highlights that in addition to the demographic aspects, great emphasis needs to be placed on the mental health of children and adolescents. The peculiarities of the relationship between health indicators show that positive indicators (*e.g.*, self-esteem) are closely, supportively, and positively related to other positive health indicators, while they are negatively correlated with negative health indicators (*e.g.*, depression). All this can be demonstrated in the present research, and adherence is related to positive health indicators in the same way and in the opposite direction to negative indicators. Accordingly, positive health indicators strengthen therapeutic collaboration, while negative indicators weaken it. For this reason, it is necessary to highlight the issue of mental hygiene and to provide support for the health of diabetic children, to support it with health prevention programs and, if necessary, intervention.

Based on our results, we could see in our study that the role of the sociodemographic variables is not significant, which can possibly be due to the characteristics of the sample. It also turned out that the relationship between the time after diagnosis and adherence was not significant, nor was the age of the diagnosis. However, we have to note that in our parallel research, the age at which the disease was diagnosed showed a significant negative effect, based on which the younger the disease is diagnosed, the higher the patient's adherence will be. Also, the adherence of patients diagnosed before the age of 12 years exceeds the adherence of patients diagnosed after the age of 12 years, and the difference between the two groups is significant.

Even if it was not in the focus of our current study, another important issue is the type of received therapy. Consistent with international research, our previous results confirm that children receiving conventional insulin therapy are characterised with significantly higher levels of depression and more negative self-esteem and emotional and behavioural symptoms perceived by the parents. This is because the results of the SEARCH research showed that the use of conventional insulin therapy (*vs* insulin pump therapy) and inadequate glycemic control (> 9%) were associated with lower quality of life in children with T1DM [43]. Regarding the development of adherence depending on the applied insulin therapy, we found that the recipients of CSII therapy scored significantly higher on the emotional feedback and peer social support subscales. It was less of a restrictive burden on them in terms of friendly relations.

Our findings suggest that a significant reduction in therapeutic adherence and metabolic control in adolescence requires significant clinical attention. It is important to emphasise that new treatment methods and approaches are needed given the complexity of the T1DM care of adolescents. In order to improve glycemic control in adolescents, it is essential to involve both parents and the healing team through a constructive and supportive system of youth autonomy that can lead to more effective diabetes self-management without violating but adhering to adolescent autonomy.

CONSENT FOR PUBLICATION

Not applicable.

CONFLICT OF INTEREST

The author declares no conflict of interest, financial or otherwise.

ACKNOWLEDGEMENTS

Declared none.

REFERENCES

[1] Williams R, Colagiuri S, Almutairi S. Diabetes Atlas. 6th ed., International Diabetes Federation 2019.

[2] Patterson CC, Karuranga S, Salpea P, *et al.* Worldwide estimates of incidence, prevalence and mortality of type 1 diabetes in children and adolescents: Results from the International Diabetes Federation Diabetes Atlas Diabetes Res Clin Pract. 9th edition. 2019; 157: p. 107842.

[3] Munkácsi B, Papp G, Felszeghy E, Nagy BE, Kovács KE. The associations between mental health, health-related quality of life and insulin pump therapy among children and adolescents with type 1 diabetes. J Pediatr Endocrinol Metab 2018; 31(10): 1065-72. [http://dx.doi.org/10.1515/jpem-2018-0130] [PMID: 30291785]

[4] Hauser ST, Jacobson AM, Lavori P, *et al.* Adherence among children and adolescents with insulin-dependent diabetes mellitus over a four-year longitudinal follow-up: II. Immediate and long-term linkages with the family milieu. J Pediatr Psychol 1990; 15(4): 527-42. [http://dx.doi.org/10.1093/jpepsy/15.4.527] [PMID: 2258799]

[5] 2010.https://www.sign.ac.uk/assets/qrg116.pdf

[6] Greening L, Stoppelbein L, Reeves CB. A model for promoting adolescents' adherence to treatment for type 1 diabetes mellitus. Child Health Care 2006; 35(3): 247-67. [http://dx.doi.org/10.1207/s15326888chc3503_4]

[7] Munkácsi B, Nagy B, Kovács KE. A diabetes adherencia vizsgálata és a Diabetes Adherencia Kérdőív Gyermek változat (DAK-GY) validálási folyamata. Orv Hetil 2019; 160(29): 1136-42. [PMID: 31303013]

[8] Aszmann A. Iskoláskorú gyermekek egészségmagatartása. Budapest: Országos Gyermekegészségügyi Intézet Nemzeti Drogmegelőzési Intézet 2003.

[9] Danckaerts M, Sonuga-Barke EJ, Banaschewski T, *et al.* The quality of life of children with attention deficit/hyperactivity disorder: a systematic review. Eur Child Adolesc Psychiatry 2010; 19(2): 83-105. [http://dx.doi.org/10.1007/s00787-009-0046-3] [PMID: 19633992]

[10] Diener E, Suh EM, Lucas RE, Smith HL. Subjective WellBeing: Three Decades of Progress. Psychol Bull 1999; 125(2): 276-302. [http://dx.doi.org/10.1037/0033-2909.125.2.276]

[11] Jekkel E, Magyar K. Az életminoség vizsgálata--áttekinto tanulmány. Psychiatr Hung 2007; 22(1): 20-32. [PMID: 17558041]

[12] Alvarado-Martel D, Velasco R, Sánchez-Hernández RM, Carrillo A, Nóvoa FJ, Wägner AM. Quality of life and type 1 diabetes: a study assessing patients' perceptions and self-management needs. Patient Prefer Adherence 2015; 9: 1315-23.
[http://dx.doi.org/10.2147/PPA.S87310] [PMID: 26396503]

[13] Agarwal AK, Singh M, Arya V, Garg U, Singh VP, Jain V. Prevalence of peripheral arterial disease in type 2 diabetes mellitus and its correlation with coronary artery disease and its risk factors. J Assoc Physicians India 2012; 60: 28-32.
[PMID: 23405538]

[14] Kiss E, Kapornai K, Baji I, Mayer L, Vetró A. Assessing quality of life: mother-child agreement in depressed and non-depressed Hungarian. Eur Child Adolesc Psychiatry 2009; 18(5): 65-273.
[http://dx.doi.org/10.1007/s00787-008-0727-3]

[15] Velő S, Keresztény Á, Miklósi M. Dallos Gy, Szentiványi D, Gádoros J, Balázs J Frissen diagnosztizált, kezelést még nem kapó figyelemhiányos-hiperaktivitás zavarú gyermekek és serdülők életminősége. Psychiatr Hung 2014; 29(4): 410-7.
[PMID: 25569830]

[16] Dallos G, Keresztény Á, Miklósi M, Gádoros J, Balázs J. Why are children and adolescents referred for psychiatric assessment without fulfiling diagnostic criteria for any psychiatric disorder? Child Adolesc Ment Health 2014; 19(3): 199-207.
[PMID: 32878372]

[17] Kiss E, Kapornai K, Baji I, Mayer L, Vetró A. Assessing quality of life: mother-child agreement in depressed and non-depressed Hungarian. Eur Child Adolesc Psychiatry 2009; 18(5): 265-73.
[http://dx.doi.org/10.1007/s00787-008-0727-3] [PMID: 19165536]

[18] Lawrence SE, Cummings EA, Pacaud D, Lynk A, Metzger DL. Managing type 1 diabetes in school: Recommendations for policy and practice. Paediatr Child Health 2015; 20(1): 35-44.
[http://dx.doi.org/10.1093/pch/20.1.35] [PMID: 25722642]

[19] Hall CA, Donza C, McGinn S, *et al.* Health-Related Quality of Life in Children With Chronic Illness Compared to Parents: A Systematic Review. Pediatr Phys Ther 2019; 31(4): 315-22.
[http://dx.doi.org/10.1097/PEP.0000000000000638] [PMID: 31568372]

[20] Vetró Á, Kiss E, Kátay L, Dombovári E. Az életminıség vizsgálata gyermek- és serdülıkorban. Psychiatr Hung 2003; 18(6): 408-17.

[21] Robitail S, Ravens-Sieberer U, Simeoni MC, *et al.* Testing the structural and cross-cultural validity of the KIDSCREEN-27 quality of life questionnaire. Qual Life Res 2007; 16(8): 1335-45.
[http://dx.doi.org/10.1007/s11136-007-9241-1] [PMID: 17668291]

[22] Jozefiak T, Larsson B, Wichstrøm L, Mattejat F, Ravens-Sieberer U. Quality of Life as reported by school children and their parents: a cross-sectional survey. Health Qual Life Outcomes 2008; 6: 34.
[http://dx.doi.org/10.1186/1477-7525-6-34] [PMID: 18489777]

[23] Dinleyici M, Çarman KB, Özdemir C, *et al.* Quality-of-life Evaluation of Healthy Siblings of Children with Chronic Illness. Balkan Med J 2019; 37(1): 34-42.
[http://dx.doi.org/10.4274/balkanmedj.galenos.2019.2019.7.142] [PMID: 31647208]

[24] Hwang U, Richardson L, Livote E, Harris B, Spencer N, Sean Morrison R. Emergency department crowding and decreased quality of pain care. Acad Emerg Med 2008; 15(12): 1248-55.
[http://dx.doi.org/10.1111/j.1553-2712.2008.00267.x] [PMID: 18945239]

[25] Cameron FJ. The impact of diabetes on health-related quality of life in children and adolescents. Pediatr Diabetes 2003; 4(3): 132-6.
[http://dx.doi.org/10.1034/j.1399-5448.2003.00019.x] [PMID: 14655271]

[26] Vanelli M, Chiarelli F, Chiari G, Tumini S. Relationship between metabolic control and quality of life in adolescents with type 1 diabetes. Report from two Italian centres for the management of diabetes in childhood. Acta Biomed 2003; 74 (Suppl. 1): 13-7.

[PMID: 12817795]

[27] Kul M, Cengel Kültür E, Senses Dinç G, Bilginer Y, Uluç S, Baykan H. Quality of life in children and adolescents with chronic kidney disease: a comparative study between different disease stages and treatment modalities. Turk J Pediatr 2013; 55(5): 493-9.
[PMID: 24382529]

[28] Tong A, Cheung KL, Nair SS, Kurella Tamura M, Craig JC, Winkelmayer WC. Thematic synthesis of qualitative studies on patient and caregiver perspectives on end-of-life care in CKD. Am J Kidney Dis 2014; 63(6): 913-27.
[http://dx.doi.org/10.1053/j.ajkd.2013.11.017] [PMID: 24411716]

[29] Moreira PI. High-sugar diets, type 2 diabetes and Alzheimer's disease. Curr Opin Clin Nutr Metab Care 2013; 16(4): 440-5.
[http://dx.doi.org/10.1097/MCO.0b013e328361c7d1] [PMID: 23657152]

[30] Pakpour AH, Zeidi IM, Hashemi F, Saffari M, Burri A. Health-related quality of life in young adult patients with rheumatoid arthritis in Iran: reliability and validity of the Persian translation of the PedsQL™ 4.0 Generic Core Scales Young Adult Version. Clin Rheumatol 2013; 32(1): 15-22.
[http://dx.doi.org/10.1007/s10067-012-2084-3] [PMID: 22965776]

[31] Almeida PHRF, Silva TBC, de Assis Acurcio F, *et al.* Quality of Life of Patients with Type 1 Diabetes Mellitus Using Insulin Analog Glargine Compared with NPH Insulin: A Systematic Review and Policy Implications. Patient 2018; 11(4): 377-89.
[http://dx.doi.org/10.1007/s40271-017-0291-3] [PMID: 29322308]

[32] Jafari P, Forouzandeh E, Bagheri Z, Karamizadeh Z, Shalileh K. Health related quality of life of Iranian children with type 1 diabetes: reliability and validity of the Persian version of the PedsQL™ Generic Core Scales and Diabetes Module. Health Qual Life Outcomes 2011; 9: 104.
[http://dx.doi.org/10.1186/1477-7525-9-104] [PMID: 22112006]

[33] Abdul-Rasoul M, AlOtaibi F, Abdulla A, Rahme Z, AlShawaf F. Quality of life of children and adolescents with type 1 diabetes in Kuwait. Med Princ Pract 2013; 22(4): 379-84.
[http://dx.doi.org/10.1159/000347052] [PMID: 23428425]

[34] Golden MP. Incorporation of quality-of-life considerations into intensive diabetes management protocols in adolescents. Diabetes Care 1998; 21(6): 885-6.
[http://dx.doi.org/10.2337/diacare.21.6.885] [PMID: 9614601]

[35] Fonagy P, Moran GS, Lindsay MK, Kurtz AB, Brown R. Psychological adjustment and diabetic control. Arch Dis Child 1987; 62(10): 1009-13.
[http://dx.doi.org/10.1136/adc.62.10.1009] [PMID: 3499865]

[36] Guttmann-Bauman I, Flaherty BP, Strugger M, McEvoy RC. Metabolic control and quality-of-life self-assessment in adolescents with IDDM. Diabetes Care 1998; 21(6): 915-8.
[http://dx.doi.org/10.2337/diacare.21.6.915] [PMID: 9614607]

[37] Delamater AM, Marrero DG, Eds. Behavioral Diabetes Social Ecological Perspectives for Pediatric and Adult Population. Genf: Springer 2020.
[http://dx.doi.org/10.1007/978-3-030-33286-0]

[38] Aldubayee M, Mohamud S, Almadani KA, *et al.* Parental levels of stress managing a child diagnosed with type 1 diabetes in Riyadh: a cross sectional study. BMC Psychiatry 2020; 20(1): 5.
[http://dx.doi.org/10.1186/s12888-019-2414-y] [PMID: 31900132]

[39] Grey M, Boland EA, Davidson M, Li J, Tamborlane WV. Coping skills training for youth with diabetes mellitus has long-lasting effects on metabolic control and quality of life. J Pediatr 2000; 137(1): 107-13.
[http://dx.doi.org/10.1067/mpd.2000.106568] [PMID: 10891831]

[40] Hoey H, Aanstoot HJ, Chiarelli F, *et al.* Good metabolic control is associated with better quality of life in 2,101 adolescents with type 1 diabetes. Diabetes Care 2001; 24(11): 1923-8.

[http://dx.doi.org/10.2337/diacare.24.11.1923] [PMID: 11679458]

[41] Graue M, Wentzel-Larsen T, Hanestad BR, Båtsvik B, Søvik O. Measuring self-reported, health-related, quality of life in adolescents with type 1 diabetes using both generic and disease-specific instruments. Acta Paediatr 2003; 92(10): 1190-6.
[http://dx.doi.org/10.1111/j.1651-2227.2003.tb02483.x] [PMID: 14632337]

[42] AlBuhairan F, Nasim M, Al Otaibi A, Shaheen NA, Al Jaser S, Al Alwan I. Health related quality of life and family impact of type 1 diabetes among adolescents in Saudi Arabia. Diabetes Res Clin Pract 2016; 114: 173-9.
[http://dx.doi.org/10.1016/j.diabres.2016.01.001] [PMID: 26830857]

[43] Naughton MJ, Ruggiero AM, Lawrence JM, *et al.* Health-related quality of life of children and adolescents with type 1 or type 2 diabetes mellitus: SEARCH for Diabetes in Youth Study. Arch Pediatr Adolesc Med 2008; 162(7): 649-57.
[http://dx.doi.org/10.1001/archpedi.162.7.649] [PMID: 18606936]

[44] Terrasson J, Terrade F, Somat A, Nivot-Adamiak S, Guitteny MA, de Kerdanet M. Association between quality of life of adolescents with type 1 diabetes and parents' illness perception as evaluated by adolescents. Psychol Health Med 2018; 23(3): 347-59.
[http://dx.doi.org/10.1080/13548506.2017.1348608] [PMID: 28675045]

[45] Angold A, Costello EJ, Worthman CM. Puberty and depression: the roles of age, pubertal status and pubertal timing. Psychol Med 1998; 28(1): 51-61.
[http://dx.doi.org/10.1017/S003329179700593X] [PMID: 9483683]

[46] Parkerson GR Jr, Connis RT, Broadhead WE, Patrick DL, Taylor TR, Tse CK. Disease-specific *versus* generic measurement of health-related quality of life in insulin-dependent diabetic patients. Med Care 1993; 31(7): 629-39.
[http://dx.doi.org/10.1097/00005650-199307000-00005] [PMID: 8326776]

[47] Dasbach EJ, Klein R, Klein BE, Moss SE. Self-rated health and mortality in people with diabetes. Am J Public Health 1994; 84(11): 1775-9.
[http://dx.doi.org/10.2105/AJPH.84.11.1775] [PMID: 7977916]

[48] Mortensen HB, Hougaard P. Comparison of metabolic control in a cross-sectional study of 2,873 children and adolescents with IDDM from 18 countries. Diabetes Care 1997; 20(5): 714-20.
[http://dx.doi.org/10.2337/diacare.20.5.714] [PMID: 9135932]

[49] Zaffani S, Maccagnan I, Morandi A, *et al.* Anxiety, Depression and Quality of Life in Italian Youths with Type 1 Diabetes Mellitus. J Diabetes Metab 2015; 6(10): 1-6.

[50] Lloyd CE, Orchard TJ. Physical and psychological well-being in adults with Type 1 diabetes. Diabetes Res Clin Pract 1999; 44(1): 9-19.
[http://dx.doi.org/10.1016/S0168-8227(99)00004-2] [PMID: 10414935]

[51] Jönsson L, Lundqvist P, Tiberg I, Hallström I. Type 1 diabetes - impact on children and parents at diagnosis and 1 year subsequent to the child's diagnosis. Scand J Caring Sci 2015; 29(1): 126-35.
[http://dx.doi.org/10.1111/scs.12140] [PMID: 24735278]

[52] Bjorner JB, Kristensen TS, Orth-Gomér K, Tibblin G, Sullivan M, Westerholm P. Self-rated health, a useful concept in research, prevention and clinical medicine. Stockholm: Forskningsrådsnämnden 1998.

[53] Viner RM, Coffey C, Mathers C, *et al.* 50-year mortality trends in children and young people: a study of 50 low-income, middle-income, and high-income countries. Lancet 2011; 377(9772): 1162-74.
[http://dx.doi.org/10.1016/S0140-6736(11)60106-2] [PMID: 21450338]

[54] Eminson DM. Medically unexplained symptoms in children and adolescents. Clin Psychol Rev 2007; 27(7): 855-71.
[http://dx.doi.org/10.1016/j.cpr.2007.07.007] [PMID: 17804131]

[55] Ursin H, Eriksen HR. The cognitive activation theory of stress. Psychoneuroendocrinology 2004;

29(5): 567-92.
[http://dx.doi.org/10.1016/S0306-4530(03)00091-X] [PMID: 15041082]

[56] Kökönyei Gy Érzelemszabályozás krónikus fejdalomban (doctoral thesis). Budapest: Eötvös Loránd Tudományegyetem, Pszichológiatudomány Doktori Iskola 2008.

[57] Salgado M, González L, Yáñez A. Parental Involvement and Life Satisfaction in Early Adolescence. Front Psychol 2021; 12: 628720.
[http://dx.doi.org/10.3389/fpsyg.2021.628720] [PMID: 33679549]

[58] Diseth TH, Tangeraas T, Reinfjell T, Bjerre A. Kidney transplantation in childhood: mental health and quality of life of children and caregivers. Pediatr Nephrol 2011; 26(10): 1881-92.
[http://dx.doi.org/10.1007/s00467-011-1887-9] [PMID: 21520007]

[59] Wisting L, Siegwarth C, Skrivarhaug T, Dahl-Jørgensen K, Rø Ø. The impact of psychological aspects, age, and BMI on eating disorder psychopathology among adult males and females with type 1 diabetes. Health Psychol Open 2020; 7(2): 2055102920975969.
[http://dx.doi.org/10.1177/2055102920975969] [PMID: 33282331]

[60] Stedman M, Rea R, Duff CJ, *et al.* People with Type Diabetes Mellitus (T1DM) self-reported views on their own condition management reveal links to potentially improved outcomes and potential areas for service improvement. Diabetes Res Clin Pract 2020; 170: 108479.
[http://dx.doi.org/10.1016/j.diabres.2020.108479] [PMID: 33002551]

[61] Edgar KA, Skinner TC. Illness representations and coping as predictors of emotional well-being in adolescents with type 1 diabetes. J Pediatr Psychol 2003; 28(7): 485-93.
[http://dx.doi.org/10.1093/jpepsy/jsg039] [PMID: 12968040]

[62] Tringer I, Varga J, Molnár E. Krónikus betegek ellátásának egészségpszichológiája.Egészségpszichológia a gyakorlatban 2014 Budapest. Medicina 2014.

[63] Leventhal H, Brissette I, Leventhal EA. The common-sense model of self-regulation of health and illness.The self-regulation of health and illness behaviour. London: Routledge 2003; pp. 42-65.

[64] Mc Sharry J, Moss-Morris R, Kendrick T. Illness perceptions and glycaemic control in diabetes: a systematic review with meta-analysis. Diabet Med 2011; 28(11): 1300-10.
[http://dx.doi.org/10.1111/j.1464-5491.2011.03298.x] [PMID: 21418098]

[65] Wisting L, Bang L, Natvig H, *et al.* Metabolic control and illness perceptions in adolescents with type 1 diabetes. J Diabetes Res 2016; 2016: 3486094.
[http://dx.doi.org/10.1155/2016/3486094] [PMID: 26682231]

[66] Law GU, Kelly TP, Huey D, Summerbell C. Self-management and well-being in adolescents with diabetes mellitus: do illness representations play a regulatory role? J Adolesc Health 2002; 31(4): 381-5.
[http://dx.doi.org/10.1016/S1054-139X(02)00397-X] [PMID: 12359385]

[67] Skinner TC, Hampson SE, Fife-Schaw C. Personality, personal model beliefs, and self-care in adolescents and young adults with Type 1 diabetes. Health Psychol 2002; 21(1): 61-70.
[http://dx.doi.org/10.1037/0278-6133.21.1.61] [PMID: 11846346]

[68] Cheng G, Yan Y. Sociodemographic, health-related, and social predictors of subjective well-being among Chinese oldest-old: a national community-based cohort study. BMC Geriatr 2021; 21(1): 124.
[http://dx.doi.org/10.1186/s12877-021-02071-7] [PMID: 33593298]

[69] Glasgow RE, McCaul KD, Schafer LC. Self-care behaviors and glycemic control in type I diabetes. J Chronic Dis 1987; 40(5): 399-412.
[http://dx.doi.org/10.1016/0021-9681(87)90173-1] [PMID: 3549758]

[70] American Psychiatric Association Diagnostic and statistical manual of mental disorders. 5th ed., Washington, DC 2013.

[71] Sørensen M, Groven KS, Gjelsvik B, Almendingen K, Garnweidner-Holme L. The roles of healthcare

professionals in diabetes care: a qualitative study in Norwegian general practice. Scand J Prim Health Care 2020; 38(1): 12-23.
[http://dx.doi.org/10.1080/02813432.2020.1714145] [PMID: 31960746]

[72] Hentinen M. Adherence to treatment. Pro Nursing Annual Book, Publication Series A1 1987; 78-82.

[73] Glasgow RE, Toobert DJ, Riddle M, Donnelly J, Mitchell DL, Calder D. Diabetes-specific social learning variables and self-care behaviors among persons with type II diabetes. Health Psychol 1989; 8(3): 285-303.
[http://dx.doi.org/10.1037/0278-6133.8.3.285] [PMID: 2767020]

[74] Wilson DP, Endres RK. Compliance with blood glucose monitoring in children with type 1 diabetes mellitus. J Pediatr 1986; 108(6): 1022-4.
[http://dx.doi.org/10.1016/S0022-3476(86)80955-6] [PMID: 3712144]

[75] Rose LE, Kim MT, Dennison CR, Hill MN. The contexts of adherence for African Americans with high blood pressure. J Adv Nurs 2000; 32(3): 587-94.
[http://dx.doi.org/10.1046/j.1365-2648.2000.01538.x] [PMID: 11012800]

[76] Burkhart P, Dunbar-Jacob J. Adherence research in the pediatric and adolescent populations: A decade in review. In: Hayman L, Mahom M, Turner R, Eds. Chronic illness in children: An evidence-based approach. New York: Springer 2002; pp. 199-229.

[77] Fotheringham MJ, Sawyer MG. Adherence to recommended medical regimens in childhood and adolescence. J Paediatr Child Health 1995; 31(2): 72-8.
[http://dx.doi.org/10.1111/j.1440-1754.1995.tb00750.x] [PMID: 7794628]

[78] Rapoff M. Adherence to pediatric medical regimens. New York: Plenum 1999.

[79] Ciechanowski PS, Katon WJ, Russo JE. Depression and diabetes: impact of depressive symptoms on adherence, function, and costs. Arch Intern Med 2000; 160(21): 3278-85.
[http://dx.doi.org/10.1001/archinte.160.21.3278] [PMID: 11088090]

[80] Horne R, Weinman J. Patients' beliefs about prescribed medicines and their role in adherence to treatment in chronic physical illness. J Psychosom Res 1999; 47(6): 555-67.
[http://dx.doi.org/10.1016/S0022-3999(99)00057-4] [PMID: 10661603]

[81] 81. Australasian Paediatric Endocrine Group for the Department of Health and Ageing. Australian Clinical Practice Guidelines, Type 1 diabetes in children and adolescents http://www.spitalmures.ro/_files/protocoale_terapeutice/endocrine/dz%20tip%20i.pdf

[82] White NH, Cleary PA, Dahms W, Goldstein D, Malone J, Tamborlane WV. Beneficial effects of intensive therapy of diabetes during adolescence: outcomes after the conclusion of the Diabetes Control and Complications Trial (DCCT). J Pediatr 2001; 139(6): 804-12.
[http://dx.doi.org/10.1067/mpd.2001.118887] [PMID: 11743505]

[83] Cameron FJ, Skinner TC, de Beaufort CE, *et al.* Are family factors universally related to metabolic outcomes in adolescents with Type 1 diabetes? Diabet Med 2008; 25(4): 463-8.
[http://dx.doi.org/10.1111/j.1464-5491.2008.02399.x] [PMID: 18294223]

[84] Court JM, Cameron FJ, Berg-Kelly K, Swift PG. Diabetes in adolescence. Pediatr Diabetes 2008; 9(3 Pt 1): 255-62.
[http://dx.doi.org/10.1111/j.1399-5448.2008.00409.x] [PMID: 18547239]

[85] Lawson ML, Sochett EB, Chait PG, Balfe JW, Daneman D. Effect of puberty on markers of glomerular hypertrophy and hypertension in IDDM. Diabetes 1996; 45(1): 51-5.
[http://dx.doi.org/10.2337/diab.45.1.51] [PMID: 8522059]

[86] Rewers M, Pihoker C, Donaghue K, Hanas R, Swift P, Klingensmith GJ. Assessment and monitoring of glycemic control in children and adolescents with diabetes. Pediatr Diabetes 2007; 8(6): 408-18.
[http://dx.doi.org/10.1111/j.1399-5448.2007.00352.x] [PMID: 18036070]

[87] Weinger K, O'Donnell KA, Ritholz MD. Adolescent views of diabetes-related parent conflict and

support: a focus group analysis. J Adolesc Health 2001; 29(5): 330-6. [http://dx.doi.org/10.1016/S1054-139X(01)00270-1] [PMID: 11691594]

[88] Anderson BJ, Svoren B, Laffel L. Initiatives to promote effective self-care skills in children and adolescents with diabetes mellitus. Dis Manag Health Outcomes 2007; 15(2): 101-8. [http://dx.doi.org/10.2165/00115677-200715020-00005]

[89] Yau M, Maclaren NK, Sperling M. 2000.https://www.ncbi.nlm.nih.gov/books/NBK498653/

[90] Lemanek KL, Kamps J, Chung NB. Empirically supported treatments in pediatric psychology: regimen adherence. J Pediatr Psychol 2001; 26(5): 253-75. [http://dx.doi.org/10.1093/jpepsy/26.5.253] [PMID: 11390568]

[91] Hippisley-Cox J, Coupland C. Development and validation of QDiabetes-2018 risk prediction algorithm to estimate future risk of type 2 diabetes: cohort study. BMJ 2017; 359: j5019. [http://dx.doi.org/10.1136/bmj.j5019] [PMID: 29158232]

[92] Petitti DB, Klingensmith GJ, Bell RA, *et al.* Glycemic control in youth with diabetes: the SEARCH for diabetes in Youth Study. J Pediatr 2009; 155(5): 668-72.e1, 3. [http://dx.doi.org/10.1016/j.jpeds.2009.05.025] [PMID: 19643434]

[93] Lustman PJ, Anderson RJ, Freedland KE, de Groot M, Carney RM, Clouse RE. Depression and poor glycemic control: a meta-analytic review of the literature. Diabetes Care 2000; 23(7): 934-42. [http://dx.doi.org/10.2337/diacare.23.7.934] [PMID: 10895843]

[94] Gandhi K, Vu BK, Eshtehardi SS, Wasserman RM, Hilliard ME. Adherence in adolescents with Type 1 diabetes: strategies and considerations for assessment in research and practice. Diabetes Manag (Lond) 2015; 5(6): 485-98. [http://dx.doi.org/10.2217/dmt.15.41] [PMID: 27066110]

[95] Markowitz JT, Garvey KC, Laffel LM. Developmental changes in the roles of patients and families in type 1 diabetes management. Curr Diabetes Rev 2015; 11(4): 231-8. [http://dx.doi.org/10.2174/1573399811666150421114146] [PMID: 25901503]

[96] Husted GR, Thorsteinsson B, Esbensen BA, *et al.* Effect of guided self-determination youth intervention integrated into outpatient visits *versus* treatment as usual on glycemic control and life skills: a randomized clinical trial in adolescents with type 1 diabetes. Trials 2014; 15: 321. [http://dx.doi.org/10.1186/1745-6215-15-321] [PMID: 25118146]

[97] Walker T. For children, skipped medications often lead to emergency department visits. Formulary 2013; 48: 313.

[98] Schwartz DD, Cline VD, Hansen JA, Axelrad ME, Anderson BJ. Early risk factors for nonadherence in pediatric type 1 diabetes: a review of the recent literature. Curr Diabetes Rev 2010; 6(3): 167-83. [http://dx.doi.org/10.2174/157339910791162952] [PMID: 20380630]

[99] Alves Peres H, Leira Pereira LR, Zangiacomine Martinez E, Viana CM, de Freitas MCF. Elucidating factors associated with non-adherence among Type 1 diabetes patients in primary care setting in Southeastern Brazil. Prim Care Diabetes 2020; 14(1): 85-92. [http://dx.doi.org/10.1016/j.pcd.2019.06.002] [PMID: 31262602]

[100] Peveler RC, Bryden KS, Neil HA, *et al.* The relationship of disordered eating habits and attitudes to clinical outcomes in young adult females with type 1 diabetes. Diabetes Care 2005; 28(1): 84-8. [http://dx.doi.org/10.2337/diacare.28.1.84] [PMID: 15616238]

[101] Gold AE, Frier BM, MacLeod KM, Deary IJ. A structural equation model for predictors of severe hypoglycaemia in patients with insulin-dependent diabetes mellitus. Diabet Med 1997; 14(4): 309-15. [http://dx.doi.org/10.1002/(SICI)1096-9136(199704)14:4<309::AID-DIA345>3.0.CO;2-#] [PMID: 9113485]

[102] Laffel LM, Connell A, Vangsness L, Goebel-Fabbri A, Mansfield A, Anderson BJ. General quality of life in youth with type 1 diabetes: relationship to patient management and diabetes-specific family conflict. Diabetes Care 2003; 26(11): 3067-73.

[http://dx.doi.org/10.2337/diacare.26.11.3067] [PMID: 14578241]

[103] Nansel TR, Weisberg-Benchell J, Wysocki T, Laffel L, Anderson B. Quality of life in children with Type 1 diabetes: a comparison of general and diabetes-specific measures and support for a unitary diabetes quality-of-life construct. Diabet Med 2008; 25(11): 1316-23. [http://dx.doi.org/10.1111/j.1464-5491.2008.02574.x] [PMID: 19046222]

[104] Paes AH, Bakker A, Soe-Agnie CJ. Impact of dosage frequency on patient compliance. Diabetes Care 1997; 20(10): 1512-7. [http://dx.doi.org/10.2337/diacare.20.10.1512] [PMID: 9314626]

[105] Dailey G, Kim MS, Lian JF. Patient compliance and persistence with antihyperglycemic drug regimens: evaluation of a medicaid patient population with type 2 diabetes mellitus. Clin Ther 2001; 23(8): 1311-20. [http://dx.doi.org/10.1016/S0149-2918(01)80110-7] [PMID: 11558867]

[106] Jarosz-Chobot P, Polanska J, Szadkowska A, *et al.* Rapid increase in the incidence of type 1 diabetes in Polish children from 1989 to 2004, and predictions for 2010 to 2025. Diabetologia 2011; 54(3): 508-15. [http://dx.doi.org/10.1007/s00125-010-1993-4] [PMID: 21165594]

[107] Weissberg-Benchell J, Antisdel-Lomaglio J. Diabetes-specific emotional distress among adolescents: feasibility, reliability, and validity of the problem areas in diabetes-teen version. Pediatr Diabetes 2011; 12(4 Pt 1): 341-4. [http://dx.doi.org/10.1111/j.1399-5448.2010.00720.x] [PMID: 21443583]

[108] Anderson B, Ho J, Brackett J, Finkelstein D, Laffel L. Parental involvement in diabetes management tasks: relationships to blood glucose monitoring adherence and metabolic control in young adolescents with insulin-dependent diabetes mellitus. J Pediatr 1997; 130(2): 257-65. [http://dx.doi.org/10.1016/S0022-3476(97)70352-4] [PMID: 9042129]

[109] Kneckt MC, Keinänen-Kiukaanniemi SM, Knuuttila ML, Syrjälä AM. Self-esteem as a characteristic of adherence to diabetes and dental self-care regimens. J Clin Periodontol 2001; 28(2): 175-80. [http://dx.doi.org/10.1034/j.1600-051x.2001.028002175.x] [PMID: 11168743]

[110] Murphy LM, Thompson RJ Jr, Morris MA. Adherence behavior among adolescents with type I insulin-dependent diabetes mellitus: the role of cognitive appraisal processes. J Pediatr Psychol 1997; 22(6): 811-25. [http://dx.doi.org/10.1093/jpepsy/22.6.811] [PMID: 9494319]

[111] Plotnikoff RC, Brez S, Hotz SB. Exercise behavior in a community sample with diabetes: understanding the determinants of exercise behavioral change. Diabetes Educ 2000; 26(3): 450-9. [http://dx.doi.org/10.1177/014572170002600312] [PMID: 11151292]

[112] Senécal C, Nouwen A, White D. Motivation and dietary self-care in adults with diabetes: are self-efficacy and autonomous self-regulation complementary or competing constructs? Health Psychol 2000; 19(5): 452-7. [http://dx.doi.org/10.1037/0278-6133.19.5.452] [PMID: 11007153]

[113] Ott J, Greening L, Palardy N, Holderby A, DeBell WK. Self-efficacy as a mediator variable for adolescents' adherence to treatment for insulin-dependent diabetes mellitus. Child Health Care 2000; 29(1): 47-63. [http://dx.doi.org/10.1207/S15326888CHC2901_4]

[114] Peyrot M, McMurry JF Jr, Kruger DF. A biopsychosocial model of glycemic control in diabetes: stress, coping and regimen adherence. J Health Soc Behav 1999; 40(2): 141-58. [http://dx.doi.org/10.2307/2676370] [PMID: 10467761]

[115] Mollema ED, Snoek FJ, Adèr HJ, Heine RJ, van der Ploeg HM. Insulin-treated diabetes patients with fear of self-injecting or fear of self-testing: psychological comorbidity and general well-being. J Psychosom Res 2001; 51(5): 665-72. [http://dx.doi.org/10.1016/S0022-3999(01)00229-X] [PMID: 11728507]

[116] Schlundt DG, Stetson BA, Plant DD. Situation taxonomy and behavioral diagnosis using prospective self-monitoring data: Application to dietary adherence in patients with type 1 diabetes. J Psychopathol Behav Assess 1999; 21: 19-36. [http://dx.doi.org/10.1023/A:1022808126697]

[117] Anderson RJ, Freedland KE, Clouse RE, Lustman PJ. The prevalence of comorbid depression in adults with diabetes: a meta-analysis. Diabetes Care 2001; 24(6): 1069-78. [http://dx.doi.org/10.2337/diacare.24.6.1069] [PMID: 11375373]

[118] de Groot M, Anderson R, Freedland KE, Clouse RE, Lustman PJ. Association of depression and diabetes complications: a meta-analysis. Psychosom Med 2001; 63(4): 619-30. [http://dx.doi.org/10.1097/00006842-200107000-00015] [PMID: 11485116]

[119] Belgrave FZ, Lewis DM. The role of social support in compliance and other health behaviors for African Americans with chronic illnesses. J Health Soc Policy 1994; 5(3-4): 55-68. [http://dx.doi.org/10.1300/J045v05n03_05] [PMID: 10138763]

[120] Schlundt DG, Sbrocco T, Bell C. Identification of high-risk situations in a behavioral weight loss program: application of the relapse prevention model. Int J Obes 1989; 13(2): 223-34. [PMID: 2744934]

[121] Karter AJ, Ferrara A, Darbinian JA, Ackerson LM, Selby JV. Self-monitoring of blood glucose: language and financial barriers in a managed care population with diabetes. Diabetes Care 2000; 23(4): 477-83. [http://dx.doi.org/10.2337/diacare.23.4.477] [PMID: 10857938]

[122] Aljasem LI, Peyrot M, Wissow L, Rubin RR. The impact of barriers and self-efficacy on self-care behaviors in type 2 diabetes. Diabetes Educ 2001; 27(3): 393-404. [http://dx.doi.org/10.1177/014572170102700309] [PMID: 11912800]

[123] Waller DA, Chipman JJ, Hardy BW, *et al.* Measuring diabetes-specific family support and its relation to metabolic control: a preliminary report. J Am Acad Child Adolesc Psychiatry 1986; 25(3): 415-8. [http://dx.doi.org/10.1016/S0002-7138(09)60265-X]

[124] Schafer LC, McCaul KD, Glasgow RE. Supportive and nonsupportive family behaviors: relationships to adherence and metabolic control in persons with type I diabetes. Diabetes Care 1986; 9(2): 179-85. [http://dx.doi.org/10.2337/diacare.9.2.179] [PMID: 3698784]

[125] Anderson BJ, Auslander WF, Jung KC, Miller JP, Santiago JV. Assessing family sharing of diabetes responsibilities. J Pediatr Psychol 1990; 15(4): 477-92. [http://dx.doi.org/10.1093/jpepsy/15.4.477] [PMID: 2258796]

[126] La Greca AM, Bearman KJ. The diabetes social support questionnaire-family version: evaluating adolescents' diabetes-specific support from family members. J Pediatr Psychol 2002; 27(8): 665-76. [http://dx.doi.org/10.1093/jpepsy/27.8.665] [PMID: 12403857]

[127] Hood KK, Butler DA, Volkening LK, Anderson BJ, Laffel LM. The Blood Glucose Monitoring Communication questionnaire: an instrument to measure affect specific to blood glucose monitoring. Diabetes Care 2004; 27(11): 2610-5. [http://dx.doi.org/10.2337/diacare.27.11.2610] [PMID: 15504994]

[128] La Greca, AM, Swales T, Klemp S, Madigan S. Self care behaviors among adolescents with diabetes. Ninth Annual Sessions of the Society of Behavioral Medicine. Baltimore, MD: Society of Behavioural Medicine p. A42 ;1988.

[129] Skinner TC, Hoey H, McGee HM, Skovlund SE. A short form of the Diabetes Quality of Life for Youth questionnaire: exploratory and confirmatory analysis in a sample of 2,077 young people with type 1 diabetes mellitus. Diabetologia 2006; 49(4): 621-8. [http://dx.doi.org/10.1007/s00125-005-0124-0] [PMID: 16525844]

[130] Iannotti RJ, Schneider S, Nansel TR, *et al.* Self-efficacy, outcome expectations, and diabetes self-management in adolescents with type 1 diabetes. J Dev Behav Pediatr 2006; 27(2): 98-105. [http://dx.doi.org/10.1097/00004703-200604000-00003] [PMID: 16682872]

[131] Hood KK, Butler DA, Anderson BJ, Laffel LMB. Updated and revised Diabetes Family Conflict Scale. Diabetes Care 2007; 30(7): 1764-9. [http://dx.doi.org/10.2337/dc06-2358] [PMID: 17372149]

[132] Cox ED, Fritz KA, Hansen KW, *et al.* Development and validation of PRISM: a survey tool to identify diabetes self-management barriers. Diabetes Res Clin Pract 2014; 104(1): 126-35. [http://dx.doi.org/10.1016/j.diabres.2014.01.015] [PMID: 24552680]

[133] Rózsa S, Várfiné Komlósi A, Kő N, Vetró Á, Gádoros J, Csorba J. A gyermek- és serdülıkori depresszió kérdıíves mérésének lehetısége a klinikai és normatív mintán szerzett tapasztalatok alapján. Pszichologia (Bp) 1999; 4: 459-82.

[134] Susánszky É, Konkoly Thege B, Stauder A, Kopp M. A WHO Jól-lét Kérdőív rövidített (WBI-5) magyar változatának validálása a Hungarostudy 2002 országos lakossági egészségfelmérés alapján. Mentálhigiéné és Psychoszomatika 2006; 7(3): 247-55.

[135] Pikó B, Barabás K, Boda K. Pszichoszomatikus tünetek epidemiológiája és hatása az egészségi állapot önértékelésére egyetemi hallgatók körében. Orv Hetil 1995; 136(31): 1667-71. [PMID: 7637988]

[136] Pikó B. Social support and health in adolescence: a factor analytical study. Br J Health Psychol 1998; 3(4): 333-44. [http://dx.doi.org/10.1111/j.2044-8287.1998.tb00578.x]

[137] Cantril H. The pattern of human concern. Rutgers University Press 1965.

[138] Gallop. Worldwide research methodology and codebook. Washington (DC): Gallup, Inc 2015.

[139] Varni JW, Seid M, Rode CA. The PedsQL: measurement model for the pediatric quality of life inventory. Med Care 1999; 37(2): 126-39. [http://dx.doi.org/10.1097/00005650-199902000-00003] [PMID: 10024117]

[140] Berkes A, Pataki I, Kiss M, *et al.* Measuring health-related quality of life in Hungarian children with heart disease: psychometric properties of the Hungarian version of the Pediatric Quality of Life Inventory 4.0 Generic Core Scales and the Cardiac Module. Health Qual Life Outcomes 2010; 8: 14. [http://dx.doi.org/10.1186/1477-7525-8-14] [PMID: 20109201]

[141] Martos T, Sallay V, Désfalvi J, Szabó T, Ittzés A. Az Élettel való Elégedettség Skála magyar változatának (SWLS-H) pszichometriai jellemzői. Mentálhigiéné és Psychoszomatika 2014; 15(3): 289-303.

[142] Goodman R. The Strengths and Difficulties Questionnaire: a research note. J Child Psychol Psychiatry 1997; 38(5): 581-6. [http://dx.doi.org/10.1111/j.1469-7610.1997.tb01545.x] [PMID: 9255702]

[143] Turi E, Gervai J, Áspán N, Halász J, Nagy P, Gádoros J. A Képességek és Nehézségek Kérdőív (SDQ-Magy) validálása serdülőkorú klinikai populációban. Psychiatr Hung 2013; 28(2): 165-79. [Validation of the Hungarian version of the Strengths and Difficulties Questionnaire in an adolescent clinical population]. [PMID: 23880514]

[144] Murillo M, Bel J, Pérez J, *et al.* Health-related quality of life (HRQOL) and its associated factors in children with Type 1 Diabetes Mellitus (T1DM). BMC Pediatr 2017; 17(1): 16. [http://dx.doi.org/10.1186/s12887-017-0788-x] [PMID: 28086765]

[145] Hood KK, Rausch JR, Dolan LM. Depressive symptoms predict change in glycemic control in adolescents with type 1 diabetes: rates, magnitude, and moderators of change. Pediatr Diabetes 2011; 12(8): 718-23. [http://dx.doi.org/10.1111/j.1399-5448.2011.00771.x] [PMID: 21564454]

[146] Goad K. Examining the Mental Health Issues Prevalent in Adolescent Type-1 Diabetics: An Evidence-Based Review to Improve Health. J Diabetes Metab 6: 599.

[147] McGrady ME, Hood KK. Cognitive–behavioral therapy for adolescents with type 1 diabetes and subclinical depressive symptoms. Diabetes Management 2013; 3(3): 207-15. [http://dx.doi.org/10.2217/dmt.13.11]

[148] de Wit M, Snoek FJ. Depressive symptoms and unmet psychological needs of Dutch youth with type 1 diabetes: results of a web-survey. Pediatr Diabetes 2011; 12(3 Pt 1): 172-6. [http://dx.doi.org/10.1111/j.1399-5448.2010.00673.x] [PMID: 20561242]

[149] Undén AL, Elofsson S, Andréasson A, Hillered E, Eriksson I, Brismar K. Gender differences in self-rated health, quality of life, quality of care, and metabolic control in patients with diabetes. Gend Med 2008; 5(2): 162-80. [http://dx.doi.org/10.1016/j.genm.2008.05.003] [PMID: 18573483]

[150] Nolen-Hoeksema S, Girgus JS. The emergence of gender differences in depression during adolescence. Psychol Bull 1994; 115(3): 424-43. [http://dx.doi.org/10.1037/0033-2909.115.3.424] [PMID: 8016286]

[151] Evans MA, Vesco A, Weissberg-Benchell JA. Depression, diabetes-related distress, and anxiety in pediatric diabetes.Behavioral Diabetes Social Ecological Perspectives for Pediatric and Adult Population. Genf: Springer 2020; pp. 49-65. [http://dx.doi.org/10.1007/978-3-030-33286-0_5]

[152] Korbel CD, Wiebe DJ, Berg CA, Palmer DL. Gender differences in adherence to type 1 diabetes management across adolescence: The mediating role of depression. Child Health Care 2007; 36: 83-98. [http://dx.doi.org/10.1080/02739610701316936]

[153] Inchley J, Currie D, Budisavljevic S. Spotlight on adolescent health and well-being Edited by: Jo Inchley, Dorothy Currie, Sanja Budisavljevic, Torbjørn Torsheim, Atle Jåstad, Alina Cosma, Colette Kelly & Ársæll Már Arnarsson Findings from the 2017/2018 Health Behaviour in School-aged Children (HBSC) survey in Europe and Canada International report VOLUME 1. KEY FINDINGS. Copenhagen: WHO Regional Office for Europe. 2017/2018.

[154] Hassan K, Loar R, Anderson BJ, Heptulla RA. The role of socioeconomic status, depression, quality of life, and glycemic control in type 1 diabetes mellitus. J Pediatr 2006; 149(4): 526-31. [http://dx.doi.org/10.1016/j.jpeds.2006.05.039] [PMID: 17011326]

[155] Williams GC, Rodin GC, Ryan RM, Grolnick WS, Deci EL. Autonomous regulation and long-term medication adherence in adult outpatients. Health Psychol 1998; 17(3): 269-76. [http://dx.doi.org/10.1037/0278-6133.17.3.269] [PMID: 9619477]

[156] Pojskic H, Eslami B. Relationship Between Obesity, Physical Activity, and Cardiorespiratory Fitness Levels in Children and Adolescents in Bosnia and Herzegovina: An Analysis of Gender Differences. Front Physiol 2018; 9: 1734. [http://dx.doi.org/10.3389/fphys.2018.01734] [PMID: 30546322]

[157] Helgeson VS, Lopez LC, Kamarck T. Peer relationships and diabetes: retrospective and ecological momentary assessment approaches. Health Psychol 2009; 28(3): 273-82. [http://dx.doi.org/10.1037/a0013784] [PMID: 19450032]

[158] Thomas AM, Peterson L, Goldstein D. Problem solving and diabetes regimen adherence by children and adolescents with IDDM in social pressure situations: a reflection of normal development. J Pediatr Psychol 1997; 22(4): 541-61. [http://dx.doi.org/10.1093/jpepsy/22.4.541] [PMID: 9302851]

[159] Baharvand P, Hormozi M. Can parents' educational level and occupation affect perceived parental support and metabolic control in adolescents with type 1 diabetes? J Educ Health Promot 2019; 8(11): 11. [PMID: 30815482]

[160] Rapoff MA, Barnard MU. Compliance with pediatric medical regimens.Patient compliance in medical

practice and clinical trials. New York: Raven Press 1991; pp. 73-98.

[161] Hood KK, Huestis S, Maher A, Butler D, Volkening L, Laffel LMB. Depressive symptoms in children and adolescents with type 1 diabetes: association with diabetes-specific characteristics. Diabetes Care 2006; 29(6): 1389-91.
[http://dx.doi.org/10.2337/dc06-0087] [PMID: 16732028]

[162] Kovacs M, Goldston D, Obrosky DS, Bonar LK. Psychiatric disorders in youths with IDDM: rates and risk factors. Diabetes Care 1997; 20(1): 36-44.
[http://dx.doi.org/10.2337/diacare.20.1.36] [PMID: 9028691]

CHAPTER 5

Recent Trials on the Cardioprotective Effects of New Generation Anti-diabetic and Lipid-Lowering Agents

Omar M. Abdelfattah[1,2], Ahmed Sayed[3], Anas Al-Refaei[3], Jasmin Abdeldayem[4], Khaled Moustafa[5], Nicholas Elias[1] and **Yehia Saleh[6,*]**

[1] *Department of Internal Medicine, Morristown Medical Center, Atlantic Health System, Morristown, New Jersey, USA*

[2] *Department of Cardiovascular Medicine, Cleveland Clinic Foundation, Cleveland, OH, USA*

[3] *Department of Internal Medicine, Faculty of Medicine, Ain Shams University, Cairo, Egypt*

[4] *Department of OB/GYN, Texas Tech University Health Sciences Center, El Paso, TX, USA*

[5] *Department of Internal Medicine, Faculty of Medicine, Alexandria University, Alexandria, Egypt*

[6] *Department of Cardiology, Houston Methodist DeBakey Heart & Vascular Center, Houston, Texas, USA*

Abstract: Diabetes and hyperlipidemia are global epidemics that significantly increase the morbidity and mortality of the affected population. Several medications have been utilized to mitigate the risk of diabetes and hyperlipidemia. Insulin, alpha-glucosidase inhibitors, thiazolidinediones have been used for decades as antidiabetic medications. Statins are a cornerstone in hyperlipidemia management. Omega□3 fatty acid supplementation has been used to treat hypertriglyceridemia with debatable effects on cardiovascular outcomes.

In the past decade, multiple new discoveries have revolutionized the management of these disorders. Sodium–glucose cotransporter 2 (SGLT2) inhibitors are a class of oral anti-diabetic drugs with a unique mechanism of action. SGLT2 was proven to reduce cardiovascular events, including hospitalization for heart failure, with this benefit extending to patients without diabetes. PCSK9 inhibitors are a new class of antihyperlipidemic that significantly lowers plasma LDL-C on top of the conventional treatment.

In this book chapter, we review the history of diabetes and hyperlipidemia medications and discuss the new classes of lipid-lowering and anti-diabetic medications and their associated cardioprotective benefits.

* **Corresponding author Yehia Saleh:** Department of Cardiology, Houston Methodist DeBakey Heart & Vascular Center, Houston, Texas, USA; Tel: +1 (517) 402-6069; E-mail: yehiatri@gmail.com

Shazia Anjum (Ed.)

Keywords: Anti-diabetic, Cardioprotective, Cardiovascular, Diabetes mellitus, DPP4, Heart failure, Incretin, Lipid, Lipid lowering, Medications, Outcomes, SGLT.

INTRODUCTION

Cardiovascular disease (CVD) is the leading cause of morbidity and mortality in patients with diabetes mellitus (DM) [1, 2]. Diabetic patients suffer from microvascular and macrovascular complications, with coronary heart disease (CHD) being the most common cardiovascular disease in diabetic patients.

Data from the UK and Canada suggest a 3% to 5% yearly decline in the rates of acute myocardial infarction (MI), stroke, cardiovascular mortality (CVM), and all-cause mortality (ACM) in patients with diabetes since the early 1990s [2]. However, patients with Type-1 Diabetes Mellitus (T1DM) and Type-2 Diabetes Mellitus (T2DM) still experience a significantly higher cardiovascular risk than the general population [2 - 5], highlighting the importance of cardioprotective medications.

Besides glycemic control, taking additional risk factors into account is also an essential component of the patient's management plan. Smoking, higher LDL levels, and a higher body mass index (BMI) are important risk factors that a holistic patient approach should consider, along with controlling plasma glucose and HbA1c levels [2].

The cardiovascular (CV) safety and benefits of novel anti-diabetic medications have been the focus of recent studies due to the heightened mortality risk in this patient population and the scarcity of evidence linking more traditional anti-diabetics to improved hard cardiovascular outcomes, such as cardiovascular mortality and major adverse cardiovascular events (including strokes and MI) [6]. This improvement in outcomes is essential because reductions in surrogate measures, such as HbA1c, are ultimately not as directly relevant to the patient as their risk of death or stroke. In addition, recent advances in lipidology have further added to the clinician's arsenal of lipid-lowering drugs, which also play a vital role in risk reduction in a variety of relatively high-risk patient populations—including those with diabetes. This chapter will discuss the newer formulations of insulin, anti-diabetic, and lipid-lowering medications and summarize each class' landmark trials, a summary of which is illustrated in Tables **1** and **2**.

Table 1. Summary of landmark clinical trials focusing on anti-diabetes medications over the past decade.

Anti-Diabetic Drugs							
SGLT2 Inhibitors							
Trial, Authors, Publication Date,	Recruitment Period	Patient Population	Intervention (Drug Name, Dose)	Control	Sample Size (Intervention vs Control)	Main Outcome(s)	
						Clinically-oriented hard outcomes	Surrogate Outcomes
EMPA-REG Trial, Zinman *et al.*, 2015 (4)	2010 to 2013	T2DM with established cardiovascular disease with an HbA1c of ≥7% and less than 10% (If they had received glucose-lowering medications within the last 12 weeks) or 9% (If they had not received such medications within the last 12 weeks)	Empagliflozin, 10 or 25mg	Placebo	4687 vs 2333	- 0.86 HR of CVM, MI or stroke - 38% RRD of CVM - 35% RRD of HF Hospitalization - 32% RRD of ACM	- An HbA1c reduction of 0.24% in the 10mg group and 0.36% in the 25mg group at 206 weeks
CANVAS, Neal *et al.*, 2017(5)	2009 to 2011 and 2014 to 2017	T2DM with an HbA1c of between 7 and 10.5% and a history of symptomatic ASCVD (Aged ≥ 30 years) or ≥2 risk factors for cardiovascular disease (Aged ≥50 years)	Canagliflozin, 100 or 300mg	Placebo	5795 vs 4347	- 0.86 HR of CVM, MI or stroke - HR of 0.73 for albuminuria progression.	- A mean HbA1c reduction of 0.58%.
CREDENCE, Perkovic *et al.*, 2019(6)	2014 to 2017	T2DM with an HbA1c of 6.5 to 12%** and albuminuric CKD (GFR of between 30 and 90ml/min/1.73m^2 in addition to an UACR of between 300 and 5000)	Canagliflozin, 100mg	Placebo	2202 vs 2199	- 0.7 HR of End-stage KD, Creatinine doubling, or CVM, or renal mortality - 0.8 HR of CVM, MI or stroke - 0.61 HR of HF Hospitalization	- Overall mean HbA1c reduction of 0.31% throughout the trial
DECLARE-TIMI 58, Wiviott *et al.*, 2019(7)	2013 to 2018	T2DM with an HbA1c of between 6.5 and 12%, a creatinine clearance of ≥60ml/min, and a history of ASCVD or multiple risk factors thereof.	Dapagliflozin, 10mg	Placebo	8582 vs 8578	- 0.83 HR of CVM or HF hospitalization - No statistically significant differences in MACE - No statistically significant differences in CVM	- Average Mean HbA1c reduction of 0.42%
DAPA-HF, McMurray *et al.*, 2019(8)	2017 to 2018	Heart Failure with NYHA class II-IV and LVEF ≤40%	Dapagliflozin, 10mg	Placebo	2373 vs 2371	- 0.74 HR of HF worsening (Hospitalization or IV therapy) or CVM - 0.82 HR of CVM - 0.83 HR of ACM - Similar effects in both diabetics and non-diabetics	- N/A
VERTIS, Cosentino *et al.*, 2020 (9)	2013 to 2019	T2DM with an HbA1c of between 7 and 10.5%, established ASCVD, and with a significant proportion of HF/EF ≤45% (23.7 and 60.7% respectively)	Ertugliflozin, 5 or 15mg	Placebo	5499 vs 2747	- RR of 0.7 for HF hospitalization - No statistically significant differences in CVM	- N/A

(Table 1) cont.....

EMPEROR, Packer *et al.*, 2020(10)	2017 to 2019	HFrEF, NYHA class II-IV	Empagliflozin, 10mg	Placebo	1863 vs 1876	– HR of 0.75 for CVM or hospitalization – HR of 0.7 for hospitalization	– N/A
Dual SGLT1/2 Inhibitors							
SOLOIST, Bhatt *et al.*, 2020(11)	2017 to 2020	T2DM recently hospitalized for worsening HF and requiring IV diuretics. (No HbA1c restrictions; median HbA1c of between 7.1 and 7.2% in the two arms respectively)	Sotagliflozin, 200 to 400mg	Placebo	608 vs 614	– HR of 0.67 for CVM, hospitalizations and urgent visits for HF. –	– Statistically insignificant difference in eGFR decline. –
SCORED, Bhatt *et al.*, 2020(12)	2018 to 2020	T2DM with an HbA1c of ≥7%, CKD (eGFR of 25 to 60 ml/min/1.73m^2), and additional CVD risk factors.	Sotagliflozin, 200 to 400mg	Placebo	5292 vs 5292	– HR of 0.74 for CVM, hospitalizations and urgent visits for HF (Amended composite endpoint) – HR of 0.84 for CVM, MI or stroke (Original coprimary endpoint) – HR of 0.77 for CVM or HF hospitalization (Original coprimary endpoint) – Statistically insignificant difference in a composite endpoint of renal transplantation, long-term dialysis, ≥ 50% reduction in the eGFR or a sustained eGFR of <15ml/min/1.73mm.	– A mean HbA1c reduction of 0.42% – Lower decline in eGFR by 1.25mL/min/1.73m^2
DDP-4 Inhibitors							
SAVOR TIMI-53, Scirica *et al.*, 2013(13)	2010 to 2011	T2DM with an HbA1c between 6.5 and 12%, and with either a history of CVD or risk factors thereof.	Saxagliptin, 5 or 2.5mg (If GFR <50)	Placebo	8280 vs 8212	– No significant difference in a composite of CVM, MI or stroke. – HR of 1.27 for HF hospitalization	– A mean HbA1c reduction of 0.2%
EXAMINE, White *et al.*, 2013(14)	2009 to 2013	T2DM with an HbA1c between 6.5 (or 7 if on insulin) and 11%, and an ACS within 15 to 90 days prior to randomization.	Alogliptin, 25mg, 12.5 (If GFR <60) or 6.25mg (If GFR <30)	Placebo	2701 vs 2679	– No increase in the risk of CVM, MI or stroke (Non-inferiority design) – Superiority not demonstrated.	– A mean HbA1c reduction of 0.36%

(Table 1) cont.....

TECOS, Green *et al.*, 2015(15)	2008 to 2012	T2DM aged ≥50 with an HbA1c of 6.5 to 8% on one to two oral hypoglycemic agents (or insulin), and with established CVD.	Sitagliptin 100mg or 50mg	Placebo	7332 vs 7332	– No significant increase in the composite end-point of CVM, MI, stroke or hospitalization for unstable angina (Non-inferiority) – Superiority not demonstrated in terms of the aforementioned end-point –	– A mean HbA1c reduction of 0.29%
MK-3102, Sheu *et al.*, 2015(16)	2010 to 2012	T2DM with an HbA1c between 7 and 10% and without a history of significant CVD.	Omarigliptin, 0.25, 1, 3, 10 or 25mg	Placebo	571 vs 114	– Hard cardiovascular outcomes not measured	– HbA1c reduction of 0.72% (25mg), 0.67% (10mg), 0.49% (3mg), 0.5% (1mg) and 0.28% (0.25mg)
MARLINA-T2D, Groop *et al.*, 2017(17)	2013 to 2015	T2DM and renal dysfunction defined as either a UACR between 30 and 3000 mg/g, or albuminuria >30mg/L of urine or >30micrograms/min	Linagliptin 5mg	Placebo	182 vs 178	– N/A	– HbA1c reduction of 0.6% – No significant change in GFR or UACR
CARMELINA, Rosenstock *et al.*, 2019(18)	2013 to 2016	T2DM with high renal risk (defined as an eGFR of 45 to 75ml/min/1.73m^2 and a UACR >200mg/g or an eGFR of 45 to 15 regardless of UACR) and high CV risk (defined as a history of disease in addition to a UACR >30mg/g	Linagliptin, 5mg	Placebo	3494 vs 3485	– No significant increase in CVM, MI, or stroke (Non-inferiority); Superiority testing insignificant – No significant increase in death due to renal failure, ESRD or 40% decline in eGFR (Non-inferiority); Superiority testing insignificant	– Mean HbA1c reduction of 0.51% –
CAROLINA, Rosenstock *et al.*, 2019 (19)	2010 to 2012	T2DM with an HbA1c between 6.5 and 8.5% (or 6.5 to 7.5%) and at high CV risk (defined as a history of existing disease or high risk thereof)	Linagliptin, 5mg	Glimepride, 1mg to 4mg	3023 vs 3010	– No significant increase in CVM, MI or stroke (Non-inferiority); Superiority testing insignificant – HR of 0.23 for hypoglycemic events	– Insignificant difference in HbA1c reductions.
RENALIS, Muskiet *et al.*, 2020(20)	2014 to 2016	T2DM with an HbA1c between 6.5 and 9%, a BMI ≥25 kg/m^2, and an eGFR >60 ml/min/1.73m^2 on metformin only.	Lingagliptin 5mg	Glimepride 1mg	23 vs 23	– N/A	– Insignificant difference in HbA1c reductions. – Increased fractional excretion of sodium and potassium did not reach statistical significance – No significant difference in GFR or UACR
GLP-1 Analogs							

(Table 1) cont.....

ELIXA, Pfeffer ***et al.*****, 2015**(21)	2010 to 2013	T2DM with an HbA1c between 5.5 and 11% whom had an ACS within 180 days.	Lixesenatide 10 to 20 micrograms	Placebo	3034 vs 3034	− No significant increase in the composite end-point of CVM, MI, stroke or hospitalization for unstable angina (Non-inferiority) − Superiority not demonstrated in terms of the aforementioned endpoint	− Mean HbA1c reduction of 0.4%
LEADER, Marso ***et al.*****, 2016**(22)	2010 to 2012	T2DM aged ≥50 with CVD or aged ≥60 with risk factors thereof.	Liraglutide 1.8mg	Placebo	4668 vs 4672	− HR of 0.78 for CVM − HR of 0.85 for ACM − No statistically significant differences in MI, stroke or hospitalizations for HF	− A mean HbA1c reduction of 0.4%.
SUSTAIN-6, Marso ***et al.*****, 2016**(23)	Feb 2013 to Dec 2013	T2DM with an HbA1c of ≥7% aged ≥50 with CVD or CKD stage III, or aged ≥60 with at least 1 CV risk factor.	Semaglutide 0.5 or 1mg	Placebo	1648 vs 1649	− HR of 0.74 for CVM, MI or stroke	− A mean HbA1c reduction of 0.7% (0.5mg) and 1% (1mg)
FREEDOM-1, Rosenstock ***et al.*****, 2017**(24)	2013 to 2014	T2DM with an HbA1c between 7.5 and 10% and 3 months or more of stable therapy (Diet/exercise and/or medications including metformin, sulfonylureas or pioglitazone)	Exenatide, 40 or 60 micrograms (automated pump)	Placebo	256 vs 154	− No data on hard outcomes	− A mean HbA1c reduction of 1%
HARMONY, Hernandez ***et al.*****, 2018**(25)	2015 to 2016	T2DM with an HbA1c of ≥7%, aged ≥40 with CVD.	Albiglutide, 30 to 50mg	Placebo	4731 vs 4732	− HR of 0.78 for CVM, MI or stroke	− A mean HbA1c reduction of 0.52%
EXSCEL, Holman ***et al.*****, 2019**(26)	2010 to 2015	T2DM with an HbA1c of 6.5 to 10% 70% of whom had CVD history	Exenatide 2mg	Placebo	7356 vs 7396	− HR of 0.86 for ACM. − No significant difference in CVM, MI or stroke (Non-inferiority) − Superiority not demonstrated in terms of the aforementioned end-point	− A mean HbA1c reduction of 0.53%
REWIND, Gerstein ***et al.*****, 2019**(27)	2011 to 2013	T2DM with an HbA1c of 9.5% or less, aged ≥50 with CVD, age ≥55 with renal dysfunction (eGFR <60ml/min/1.73m^2) and vascular pathology, or age ≥60 with CV risk factors.	Dulaglutide 1.5mg	Placebo	4949 vs 4952	− HR of 0.88 for CVM, MI or stroke − No statistically significant difference with respect to ACM	− A mean HbA1c reduction of 0.61%

(Table 1) cont.....

PIONEER-6, Husain *et al.*, 2019(28)	Jan 2017 to Aug 2017	T2DM aged ≥50 years with CVD or CKD, or ≥60 with CV risk factors only; No HbA1c restrictions, with an average of 8.2% in both arms.	Semaglutide, target dose of 14mg	Placebo	1591 vs 1592	– HR of 0.49 for CVM – HR of 0.51 for ACM – No significant difference in a composite of CVM, MI or stroke (Non-inferiority) – Superiority not demonstrated in terms of the aforementioned end-point	– A mean HbA1c reduction of 0.7%
Insulin							
Origin, The ORIGIN investigators, 2012(29)	2003 to 2011	Adults aged ≥50 with CV risk factors and with early T2DM, impaired fasting glucose or impaired glucose tolerance.	Glargine, dose titrated to a target FBG of 95mg/dl	Standard of care	6264 vs 6273	– No difference in CVM, MI, stroke, revascularization or HF hospitalization – Reduced onset of new diabetes – Increased rates of hypoglycemia	– A mean HbA1c reduction of 0.3%
DEVOTE-7, Marso *et al.*, 2017(30)	2013 to 2014	T2DM with an HbA1c of ≥7% (or less if using ≥20 basal insulin units), aged ≥50 years with CVD or CKD, or ≥60 with CV risk factors only.	Degludec, dose titrated to a glucose target of 71 to 90mg before breakfast	Glargine, dose titrated to a glucose target of 71 to 90mg before breakfast	3818 vs 3819	– No significant difference in CVM, MI or stroke (Non-inferiority) – RR of 0.6 for hypoglycemic events	– No significant difference in HbA1c – Significantly lower fasting glucose levels
CONCLUDE, Philis-Tsimikas *et al.*, 2019(31)	2017 to 2019	T2DM with an HbA1c of ≤9.5% treated with basal insulin at risk for hypoglycemia.	Degludec 200 U	Glargine 300 U	805 vs 804	– No statistically significant difference in the overall rates of symptomatic hypoglycemia – Significant reduction in the rates of nocturnal symptomatic and severe hypoglycemia	– Insignificant difference in HbA1c reductions.
ACHIEVE, Meneghini *et al.*, 2020(32)	2015 to 2018	Insulin-naïve T2DM with an HbA1c ≥8% despite ≥2 antihyperglycemics for ≥1 year	Glargine 300 U/mL	Glargine 100U/mL or Detemir	1651 vs 1653	– No data on hard outcomes	– OR of 1.19 for meeting HbA1c targets without clinically significant (level 2) hypoglycemic events – Insignificant difference in HbA1c reductions.
ORMD-0801, Eldor *et al.*, 2020(33)	2015 to 2016	T2DM with an HbA1c of ≥7.5% uncontrolled by multiple oral medications	ORMD-0801, 8 or 16 or 32mg	Placebo	354 (total)	– No significant? difference in hypoglycemic events	– Significant reductions in HbA1c levels

(Table 1) cont.....

Icodec, Rosenstock *et al.*, 2020(34)	2018 to 2019	T2DM with an HbA1c of between 7 and 9.5% on metformin with no prior long-term insulin therapy.	Icodec once-weekly, starting dose of 70 U adjusted to achieve adequate control	Glargine once-daily, starting dose of 10 U adjusted to achieve adequate control	125 vs 122	– No data on hard outcomes	– No statistically significant reduction in HbA1c levels – Significantly higher incidence of hypoglycemic alerts, but no increase in the incidence of clinically significant or severe hypoglycemia.
Alpha-glucosidase inhibitors							
ACE, Holman *et al.*, 2017 (35)	2009 to 2015	Patients aged ≥50 with CHD, impaired glucose tolerance and 80% adherence during the run-in period.	Acarbose 150mg	Placebo	3272 vs 3250	– No statistically significant differences in CVM, MI, stroke, or admissions for unstable angina or HF	– Statistically significant reduction in HbA1c levels by 0.06% – HR of 0.82 for the development of new-onset diabetes
Thiazolidinediones							
PROactive, Dormandy *et al.*, 2005 (35)	2004 to 2005	T2DM with an HbA1c ≥6.5% and extensive macrovascular disease (defined as an MI/Stroke within 3 months, an ACS within 6 months, CAD or lower limb arterial disease)	Pioglitazone, 15 to 45mg	Placebo	2605 vs 2633	– No significant difference in the primary composite endpoint of ACM, MI, stroke, ACS, surgical/endovascular intervention in the legs/coronaries, or amputations – HR of 0.84 for ACM, MI or stroke	– A mean HbA1c reduction of 0.5%
RECORD, Home *et al.*, 2009 (36)	2001 to 2003	T2DM on maximally tolerated metformin/sulfonylureas monotherapy and a BMI ≥25kg/m^2 without a history of hospitalization for CV events prior to enrollment and without heart failure.	Rosiglitazone, 4 to 8mg or to target of <7% HbA1c + Metformin or sulfonylurea	Metformin and sulfonylurea	321 vs 323	– No significant differences in the composite of CVM or CV Hospitalizations (Non-inferiority) – HR of 2.1 for HF-related death or hospitalization – Increased risk of fractures	– A mean HbA1c reduction of 0.27% (background of metformin) or 0.26% (background of sulfonylurea)

Abbreviations: *HR, Hazard Ratio; CVM, Cardiovascular Mortality; MI, Myocardial Infarction; RRD, Relative Risk Reduction; ACM, All-cause mortality; HF, Heart Failure; ACS, Acute Coronary Syndrome; BMI, Body Mass Index; CKD, Chronic Kidney Disease; GFR, Glomerular Filtration Rate;EMPAREG, Empagliflozin Cardiovascular Outcome Event Trial in Type 2 diabetes Mellitus Patients; CANVAS, CANagliflozin cardioVascular Assessment Study; CREDENCE, Evaluation of the Effects of Canagliflozin on Renal and Cardiovascular Outcomes in Participants With Diabetic Nephropathy; DM-TIMI 58, The Dapagliflozin Effect on Cardiovascular Events–Thrombolysis in Myocardial Infarction 58; DAPA-HF, Dapagliflozin and Prevention of Adverse Outcomes in Heart Failure; VERTIS, eValuation of ERTugliflozin efficacy and Safety CardioVascular outcomes; EMPEROR, EMPagliflozin outcomE tRial in Patients With chrOnic heaRt Failure With Reduced Ejection Fraction; SOLOIST, Effect of Sotagliflozin on Cardiovascular Events in Patients With Type 2 Diabetes Post Worsening Heart Failure; SCORED, Effect of Sotagliflozin on Cardiovascular and Renal Events in Patients With Type 2 Diabetes and Moderate Renal Impairment Who Are at Cardiovascular Risk; SAVOR-TIMI-53, The investigators in the Saxagliptin Assessment of Vascular Outcomes Recorded in Patients with Diabetes Mellitus–Thrombolysis in Myocardial Infarction-53; EXAMINE, The Examination of Cardiovascular Outcomes With Alogliptin vs Standard of Care in patients with type 2 diabetes mellitus and acute coronary syndrome; MARLINA-T2D, Efficacy, Safety and Modification of Albuminuria in Type 2 Diabetes Subjects with Renal Disease with LINAgliptin;*

CARMELINA, The CV and kidney clinical profile of linagliptin is being comprehensively evaluated in the CArdiovascular safety and Renal Microvascular outcomE study with LINAgliptin; CAROLINA, CARdiovascular Outcome study of LINAgliptin versus glimepiride in patients with type 2 diabetes; RENALIS, Renal Effects of DPP-4 Inhibitor Linagliptin in Type 2 Diabetes; ELIXA, Evaluation of Cardiovascular Outcomes in Patients With Type 2 Diabetes After Acute Coronary Syndrome During Treatment With AVE0010 (Lixisenatide); TECOS, Assessing the Safety of Sitagliptin in Older Participants in the Trial Evaluating Cardiovascular Outcomes with Sitagliptin; LEADER, Liraglutide Effect and Action in Diabetes: Evaluation of Cardiovascular Outcome; SUSTAIN-6, Trial to Evaluate Cardiovascular and Other Long-term Outcomes with Semaglutide in Subjects with Type 2 Diabetes; FREEDOM-1, Efficacy and Safety of ITCA 650, a Novel Drug-Device GLP-1 Receptor Agonist, in Type 2 Diabetes Uncontrolled With Oral Antidiabetes Drugs; HARMONY, Albiglutide and cardiovascular outcomes in patients with type 2 diabetes and cardiovascular disease; EXSCEL, The EXenatide Study of Cardiovascular Event Lowering; REWIND, Effect of Dulaglutide on Major Cardiovascular Events in Patients With Type 2 Diabetes: Researching Cardiovascular Events With a Weekly INcretin in Diabetes; PIONEER-6, Trial Investigating the Cardiovascular Safety of Oral Semaglutide in Subjects With Type 2 Diabetes; ORIGIN, Outcome Reduction With Initial Glargine Intervention; DEVOTE-7, Trial Comparing Cardiovascular Safety of Insulin Degludec *versus* Insulin Glargine in Subjects With Type 2 Diabetes at High Risk of Cardiovascular Events; CONCLUDE, Trial Comparing the Efficacy and Safety of Insulin Degludec and Insulin Glargine 300 Units/mL in Subjects With Type 2 Diabetes Mellitus Inadequately Treated With Basal Insulin With or Without Oral Antidiabetic Drugs; ACE, Effects of acarbose on cardiovascular and diabetes outcomes in patients with coronary heart disease and impaired glucose tolerance; PROactive, Prospective Pioglitazone Clinical Trial In Macrovascular Events; RECORD, Rosiglitazone Evaluated For Cardiovascular Outcomes.
**The trial endpoint was amended because of a loss of funding and a lower-than-expected event rate.*
**Amended to an upper limit of 10.5% in Germany.

Table 2. Summary of landmark clinical Trials focusing lipid-lowering drugs over the past decade.

Lipid-Lowering Drugs							
PCSK9-Inhibitors							
Trial, Authors, Publication Date	**Recruitment Period**	**Patient Population**	**Intervention (Drug Name & Dose)**	**Control**	**Sample Size (Intervention vs control)**	**Main Outcome(s)**	
						Clinically-oriented hard outcomes	**Surrogate Outcomes**
GAUSS, Sullivan *et al.*, 2012(37)	2011 to 2012	Adults with hypercholesterolemia and statin intolerance to any drug within the class.	Evolocumab, 280, 350 or 420mg with or without Ezetemibe 10mg	Ezetimibe 10mg + Placebo	125 vs 32	– No data on hard outcomes	– LDL reduction of 47.3 mg/dL (420mg + Ezetimibe), 35.9 mg/dL (420mg), 27.8 (350mg) and 26 mg/dL (280mg)
RUTHERFORD-2, Raal *et al.*, 2014(38)	2011 to 2012	Patients with familial hypercholesterolemia	Evolocumab, 140 or 420mg	Placebo	221 vs 110	– No data on hard outcomes	– LDL reduction of 59.2% (140mg) and 61.3% (420mg) –
GAUSS-2, Stroes *et al.*, 2014(39)	Jan 2013 to Nov 2013	Adults with hypercholesterolemia and statin intolerance to at least 2 different drugs.	Evolocumab, 140mg or 420mg	Ezetimibe 10mg	205 vs 102	– No data on hard outcomes	– Significantly reduced LDL levels by 38.1% (140mg) and 37.6% (420mg) –
DESCARTES, Blom *et al.*, 2014(40)	2012 to 2013	Patients with hypercholesterolemia and triglycerides ≤400mg/dL	Evolocumab, 140 or 420mg	Placebo	599 vs 302	– No data on hard outcomes	– LDL reduction of 36.9 and 38.7% respectively

(Table 2) cont.....

MENDEL-2, Koren *et al.*, 2014(41)	Jan 2013 to Oct 2013	Adults with hypercholesterolemia and a a 10-year Framingham risk score of ≤10%	Evolocumab, 140mg or 420mg	Placebo or Ezetimibe	256 vs 308	– No data on hard outcomes	– Significantly reduced LDL levels by 55 to 57% (vs placebo) and 38 to 40% (vs Ezetemibe)
LAPLACE-2, Robinson *et al.*, 2014(42)	Jan 2013 to Nov 2013	Incompletely controlled hypercholesterolemia (No statin usage and High LDL or Statin usage insufficient to meet targets)	Evolocumab, 140 or 420mg	Ezetemibe 10mg or Placebo	1117 vs 779	– No data on hard outcomes	– Significantly reduced LDL levels by 66 to 75% (140mg) and 63 to 75% (420mg) vs placebo, as well as 37.5 to 44.9% (140mg) and 43.5 to 43.8% (420mg) vs ezetimibe.
YUKAWA-1, Hirayama *et al.*, 2014(43)	2012 to 2013	High-risk patients stable on statins with or without ezetimibe (High-risk defined as a history of ASCVD, familial hypercholesterolemia, T2DM ≥3months prior to randomization or the presence of ≥3 additional risk factors)	Evolocumab, 70, 140, 280 or 420mg	Placebo	205 vs 102	– No data on hard outcomes	– LDL reduction of 68.6 and 63.9% (140mg Q2W and 420mg QM respectively)
SPIRE, Ridker *et al.*, 2017(44)	2013 to 2017	Patients with a previous CV event or at high risk thereof (high risk defined as a history of diabetes, CKD, PVD with additional risk factors, or familial hypercholesterolemia)	Bococizumab, 150mg	Placebo	13720 vs 13718	– HR of 0.79 for CVM, MI, stroke, or unstable angina requiring hospitalization and urgent revascularization in the high-risk arm (LDL > 100mg/dl) – No significant benefit in low-risk patients (LDL > 70mg/dl) for the same composite outcome	– Significant LDL reduction by 59%
FOURIER, Sabatine *et al.*, 2017(45)	2013 to 2016	Patients with a history of ASCVD and an LDL ≥70mg/dL despite statin therapy.	Evolocumab, 140 or 420mg	Placebo	13784 vs 13780	– HR of 0.85 for CVM, MI, stroke, unstable angina requiring hospitalization, or coronary revascularization – HR of 0.8 for CVM, MI or stroke	– Significant LDL reduction by 54%.
ODYSSEY, Szarek *et al.*, 2019(46)	2012 to 2018	Patients with a history of ACS and an LDL ≥70mg/dL despite statin therapy	Alirocumab 75mg	Placebo	9462 vs 9462	– HR of 0.83 for CVM – HR of 0.87 for non-fatal events (Stroke, MI, revascularization or hospitalization for unstable angina or HF)	– N/A

(Table 2) cont.....

OSLER-1, Koren *et al.*, 2019(47)	2011 to 2018	Patients from LAPLACE-TIMI 57, MENDEL-1, YUKAWA-1, RUTHERFORD-1 or GAUSS-1 continuing therapy.	Evolocumab (Dose dependent on original trial)	Placebo	1255 vs 442	– No data on hard outcomes	– LDL reduction of 56%
Omega-3 Fatty Acids							
OMEGA, Rauch *et al.*, 2010	2003 to 2008	Patients aged ≥18 admitted due to acute STEMI or non-STEMI	1g of Omega-3 fatty acids (EPA 460mg/DHA 380mg)	Placebo	1925 vs 1893	– No improvement in all-cause mortality, , MACE or sudden cardiac death, revascularization,	– N/A
ORIGIN, ORIGIN investigators, 2012(48)	2003 to 2005	Patients aged ≥50 with diabetes (on no more than one oral hypoglycemic) or who have impaired glucose tolerance who have a history of (or who are at high risk for) CVD.	1g of Omega-3 fatty acids of ≥90% purity (EPA 465mg/DHA 375mg)	Placebo (1g of olive oil)	6281 vs 6255	– No significant benefit in ACM, CVM, arrhythmias or major vascular events	– Trigylceride reduction of 14.5mg/dL
ASCEND, ASCEND investigators, 2018(49)	2005 to 2011	Patients age ≥40 with diabetes and no CVD.	1-g capsules of 840mg of omega-3 fatty acids (EPA 460mg/DHA 380mg)	Placebo (1g of olive oil)	7740 vs 7740	– No significant benefit in ACM, revascularization or serious vascular events.	– N/A
REDUCE-IT, Bhatt *et al.*, 2019(50)	2011 to 2018	Patients with a triglyceride level of between 150 and 499 mg/dL age ≥45 with established CVD or ≥50 with DM and ≥1 additional risk factors.	Icosapent Ethyl, 4mg	Placebo	4089 vs 4090	– HR of 0.75 in the primary composite of CVM, MI, stroke, revascularization or unstable angina – HR of 0.74 in the secondary composite of CVM, MI or stroke	– Median reduction of triglycerides increased by 19.7%. – Decreased the rise in LDL levels by 6.6%
STRENGTH, Nicholls *et al.*, 2020(51)	2014 to 2020	Patients with established CVD, patients with T1DM or T2DM aged ≥ 40 if men and ≥50 if women who had at least one additional risk factor, or patients aged ≥50 if men and ≥60 if women with at least one additional risk factor. All participants were required to have triglyceride levels between 180 and 500 mg/dL	4g of omega-3 (Containing EPA and DHA)	Placebo (Corn-oil)	6539 vs 6539	– Significant reduction in triglycerides by 18.1% – 5% reduction in non-HDL cholesterol.	– No significant improvement in the primary endpoint of CVM, MI, stroke, revascularization or hospitalization for unstable angina. – 10% Higher rate of gastrointestinal adverse effects.
TRILOGY-1,	2018 to 2019	Patients with fasting triglycerides between 500 and 1500 mg/dL.	CaPre (Omega---phospholipid)	Placebo		–	–

Abbreviations: *HR, Hazard Ratio; CVM, Cardiovascular Mortality; MI, Myocardial Infarction; HF, Heart Failure; PCSK9, Proprotein convertase subtilisin/kexin type 9; DM, Diabetes Mellitus; EPA, eicosapentaenoic acid; DHA, docosahexaenoic acid ; GAUSS, Evolocumab vs. Ezetimibe in Statin-Intolerant Hyperlipidemic Japanese Patients; RUTHERFORD, the reduction of LDL-C with PCSK9 inhibition in heterozygous familial hypercholesterolemia disorder randomized trial; DESCARTES, Durable Effect of PCSK9 Antibody CompARed wiTh placEbo Study; MENDEL, Monoclonal Antibody Against PCSK9 to Reduce Elevated Low-density Lipoprotein Cholesterol (LDL-C) in Adults Currently Not Receiving Drug*

Therapy for Easing Lipid Levels; LAPLACE, LDL-C Assessment With PCSK9 Monoclonal Antibody Inhibition Combined With Statin Therapy-2; YUKAWA, A Study to Evaluate Tolerability and Efficacy of Evolocumab (AMG 145) in Japanese Subjects; SPIRE-2, The Evaluation of Bococizumab in Reducing the Occurrence of Major Cardiovascular Events in High Risk Subjects; FOURIER, Further Cardiovascular Outcomes Research With PCSK9 Inhibition in Subjects With Elevated Risk; ODYSSEY, Evaluation of Cardiovascular Outcomes After an Acute Coronary Syndrome During Treatment With Alirocumab; OSLER-1, Open Label Study of Long Term Evaluation Against LDL-C Trial; OMEGA, Effect of Omega 3-Fatty Acids on the Reduction of Sudden Cardiac Death After Myocardial Infarction; ORIGIN, Outcome Reduction With Initial Glargine Intervention; ASCEND, A Study of Cardiovascular Events iN Diabetes; REDUCE-IT, Evaluation of the Effect of AMR101 on Cardiovascular Health and Mortality in Hypertriglyceridemic Patients With Cardiovascular Disease or at High Risk for Cardiovascular Disease; STRENGTH, Outcomes Study to Assess STatin Residual Risk Reduction With EpaNova in HiGh CV Risk PatienTs With Hypertriglyceridemia; TRILOGY, A Phase 3 STudy of CaPRe In LOwering Very hiGh TriglYcerides.

SODIUM/GLUCOSE COTRANSPORTER-2 INHIBITORS (SGLT2I)

History, Mechanisms, and Pleiotropic Effects of SGLT2i

The Sodium/glucose cotransporter-2 inhibitors (SGLT2i) are a novel type of glucose-lowering drugs that have recently attained a significant role in the management of T2DM in addition to a recent expansion in clinical utilization that saw them gain ground in treating heart failure. The FDA approved four drugs of this class, namely canagliflozin, the first-in-class drug to be approved in 2013, and, subsequently, emapagliflozin, dapagliflozin, and ertugliflozin. At the time of writing, sotagliflozin, though approved in Europe, has not been yet approved by the FDA, but the results of recently published clinical trials (see below) may change that.

Sodium-glucose cotransporters are found in both the gastrointestinal system, where they are needed for the absorption of glucose, and in the renal tubules, where they are needed for glucose reabsorption. The initial inspiration for SGLT2i was derived from dihydrochalcone phlorizin, a substance obtained from trees that had been in experimental use for over a century. Phlorizin is a nonselective SGLT2i that inhibits both enteric absorption and renal reabsorption of glucose, but it, unfortunately, has a relatively low bioavailability owing to its metabolism by intestinal enzymes, which in large part explains the delay seen between its discovery and the use of modern analogs in the practice of medicine [55, 56].

SGLT2i drugs compete with glucose for the aforementioned transporters, which results in a reduced rate of glucose reabsorption and increased excretion in the urine, coupled with a complementary decrease in sodium reabsorption that may also play an essential role in endowing this class of drugs with cardiovascular benefits. Moreover, the increase in glucose excretion may essentially create an energy deficit that facilitates weight loss, which can contribute to the observed cardiovascular benefits [55]. Furthermore, SGLT2i, by acting on specific electrolyte transporters within the cardiac muscle, may improve myocardial energy metabolism and slow cardiac remodeling [55].

Pivotal Clinical Trials Involving SLGT2i

The EMPA-REG trial [7] represented a breakthrough in the treatment of diabetes in that it was the first trial to demonstrate a significant reduction in clinical outcomes with the use of SGLT2i. The trial randomized 7,028 patients with DM and established CVD between 2010 and 2013, with a median follow-up time of 3.1 years. The primary outcome was CVM, MI, or stroke, with the secondary outcome incorporating the three previous endpoints plus unstable angina (UA) hospitalization. The composite primary outcome, ACM and HF hospitalizations, were significantly lower in the empagliflozin group (RRR of 14, 32, and 35%, respectively), with similar efficacy being shown at both 10 and 25 mg dosages. The trial also played a pivotal role in the FDA's 2016 approval of empagliflozin for the treatment of diabetes, and, in many ways, it set the stage for an oncoming wave of clinical trials that would change the landscape of diabetes treatment for years to come.

The CANVAS study consisted of 2 sister trials that aimed to assess canagliflozin's cardiovascular benefits in high-risk T2DM [8]. There were 10,142 patients in the CANVAS program, 4,330 patients in the CANVAS trial, and 5,812 in the CANVAS-R trial, with a combined analysis being conducted to maximize statistical power. The mean follow-up duration was 188.2 weeks, with a mean DM history duration of 13.5 years. The primary outcome was CVM, MI, or stroke, and the secondary outcomes were ACM, CVM, and HF hospitalization. Canagliflozin significantly improved the primary composite endpoint (RRD of 14%). Though the endpoints of ACM and CVM did not reach statistical significance, fewer patients assigned to canagliflozin were hospitalized for heart failure. The rationale behind designing the sister trial (CANVAS-R) was assessing the renal-protection bestowed by canagliflozin (if any); and indeed, albuminuria progression was significantly less likely in the canagliflozin arm.

However, these improvements did not come free of cost, with the canagliflozin arm having a higher rate of amputations, a finding that prompted caution when prescribing what was –at the time—a relatively new class of medications. Interestingly, though not the first trial to be published in full, it led to the first-in-class FDA approval in 2013, as the trial was initiated earlier than EMPA-REG and was approved based on interim results.

The next trial to test canagliflozin's efficacy was CREDENCE, which assessed the effects of the SGLT2 inhibitor on renal outcomes in 4401 patients with T2DM and albuminuric chronic kidney disease [9]. The primary outcome was a composite of renal mortality, CVM, end-stage kidney disease, or creatinine doubling, whereas the secondary outcomes were specified in the following order: first, a composite of cardiovascular mortality or heart failure hospitalization; second, a composite of cardiovascular death, myocardial infarction, or stroke; third, heart failure hospitalizations. Canagliflozin significantly improved the primary endpoint (HR of 0.7; NNT of 24), the secondary endpoint of CVM, MI, or stroke (HR of 0.8; NNT of 43), and HF hospitalization (HR of 0.61; NNT of 43).

The significance of CREDENCE was that it was the first beneficial trial dedicated to patients with T2DM and concomitant kidney disease, which is an invaluable finding as renal affection in diabetes portends a grimmer prognosis and accelerates the already deteriorating cardiovascular system, putting these patients at a particularly high risk of CV events.

Unfortunately, the next major trial, DECLARE-TIMI-58 [10], did not yield results that were as impressive as its predecessors. It evaluated the impact of 10 mg of dapagliflozin on renal and cardiac outcomes in a high-risk T2DM population. Randomization of 17,160 patients was done, 40.6% of whom had established atherosclerotic disease and a median follow-up of 4.2 years. The primary safety outcome was CVM, MI, or stroke, along with hospitalization for HF (which was added half-way through the trial due to the findings of the EMPA-REG trial mentioned above). Dapagliflozin reduced the composite of CVM and HF hospitalization (HR of 0.83; NNT of 111); furthermore, renal events (decline in eGFR, new ESKD, or death due to renal or cardiovascular causes) were also less likely in the dapagliflozin group (RRR of 24%; NNT of 77).

However, there was no statistically significant reduction in the rate of the primary outcome of CVM, MI, or stroke (it was merely non-inferior to placebo), and thus did not represent as much of a breakthrough as previous trials did. However, these relatively disappointing results were merely a minor bump on the road for dapagliflozin, as the next major trial, DAPA-HF, broadened the usage of SGLT2i to an entirely new patient population.

DAPA-HF [11] was a trial to evaluate the efficacy and safety of 10 mg of dapagliflozin in patients with heart failure and a reduced ejection fraction, regardless of the presence of diabetes. 4,744 patients with NYHA class II-IV and an ejection fraction <40% were randomized in this trial with a follow-up of 18.2 months. The primary endpoint was a dual-composite of worsening heart failure or death from cardiovascular causes, whereas the secondary endpoint was a composite of heart failure hospitalization or cardiovascular mortality.

Dapagliflozin significantly reduced the primary outcome (RRR of 26%; NNT of 21) as well as the secondary outcome (RRR of 25%; NNT of 125). More impressively, the incidence of CVM (RRR of 0.82; NNT of 53) and ACM (RRR of 0.83; NNT of 44) was also significantly reduced in the dapagliflozin arm.

The benefits of dapagliflozin extended to both patients with and without diabetes, thus dissipating worries that it may only be useful in select patient populations. The importance of DAPA-HF cannot be understated: not only did it demonstrate benefit in a new patient population, but it did so by showing a reduction in the hardest of all outcomes: mortality (even when taken as a single endpoint in and of itself). These findings played a significant role in the FDA's decision to approve dapagliflozin's use in heart failure in 2020, further diversifying the armamentarium of physicians in treating what both a relatively common and highly fatal condition is.

The next SGLT2i to be trialed was ertugliflozin, with the VERTIS trial aiming to compare two doses of ertugliflozin (5 and 15 mg) *versus* placebo in a population composed of high-risk patients with T2DM, accruing 8,246 patients over six years, with a large portion of patients having HF and/or a reduced ejection fraction. Overall, the trial's results were somewhat of a disappointment, as the drug resulted in an overall decrease in HF hospitalization (RRR of 30%; NNT of 91), but there was no statistically significant difference in CVM.

Not to be dampened by ertugliflozin's disappointing results, the EMPEROR-Reduced trial, which was only the second trial after DAPA-HF to target HF, recruited slightly over 3,700 patients and assigned them 10 mg of empagliflozin *versus* placebo. Like DAPA-HF, the trial only included patients with a reduced ejection fraction, and a significant improvement in a composite of CVM and hospitalization was seen (RRR of 25%; NNT of 19), though it is worth noting that most of the composite benefit was derived from a reduction in hospitalizations rather than mortality. This stands in sharp contrast to DAPA-HF, where a statistically significant reduction in mortality was observed. Nevertheless, being only the second large-scale randomized trial to demonstrate a benefit for an SGLT2i in this patient population, this trial's results will play a key role in cementing the importance of SGLT2i in HF for many years to come.

Dual SGLT1/2 Inhibitors: The Next Step in the Evolution of SGLT2i?

The next generation of SGLT2i to be trialed were dual SGLT1/2 inhibitors, which have a theoretical edge over traditional SGLT2i in that traditional SGLT2i essentially allow the kidney to compensate for SGLT2 inhibition by increasing the activity of SGLT1; therefore, less glucose is excreted than would typically be expected due to the upregulation of SGLT1. By blocking both transporters simultaneously, one might be able to achieve a greater degree of glucose excretion and, by extension, improved clinical outcomes [57]. Two recently published 2020 trials, both by Bhatt *et al.*, sought to investigate their efficacy in two different patient populations, though the trials were hampered by premature termination [14, 15].

The SOLOIST trial [14], which included T2DM patients recently hospitalized for HF and requiring diuretic therapy, had a relatively small sample size of just 1,222 patients. Patients were subsequently randomized into the sotagliflozin arm (with a dose ranging from 200 to 400 mg) or the placebo arm. Nevertheless, despite the somewhat small sample, the trial results showed a significant reduction in the composite outcome of CVM, hospitalization, and urgent HF visits (RRR of 33%; NNT of 4); disappointingly, the trial did not seem to show significant renal-protective effects as there was no significant difference between the two groups in terms of the decline in estimated GFR.

The other trial, SCORED [58], which studied the same intervention, enrolled high-risk T2DM patients with chronic kidney disease (CKD), defined as an eGFR of between 25 to 60 ml/min/1.73m^2. As a result of the premature halting of the trial and the lower-than-expected event-rate, the two original primary endpoints were amended. Both original endpoints, the first being a composite of CVM, MI, or stroke and the second being a composite of CVM or HF hospitalization, showed an impressive benefit in the sotagliflozin arm (RRR of 16% and 23%, respectively). The amended endpoint, a composite of CVM, HF hospitalization, or urgent visits, was similarly significantly reduced in the sotagliflozin group (RRR of 26%).

Also, unlike the SOLOIST trial, which found no evidence of improved renal benefits, the SCORED trial reduced the decline of the eGFR by 1.25ml/min/1.73m^2. Unfortunately, the more challenging clinical endpoints of renal transplantation, long-term dialysis, ≥50% eGFR reduction, or a sustained eGFR <15ml/min/1.73mm showed no statistically significant differences.

Overall, the addition of SGLT2i to the physician's arsenal has revolutionized the landscape of diabetes treatment, with improvements in cardiovascular outcomes being perhaps the single most significant advantage of this class of drugs. SGLT2i have gained prominence in recent years and are recommended for diabetic patients at high-risk for CVD regardless of glycemic control, which is reflective of a broader move towards more clinically relevant, patient-oriented outcomes as opposed to merely aiming at keeping glucose levels in check. This has culminated in this class of drugs featuring heavily in the 2018 ACC guidelines on CV risk reduction in patients with diabetes and/or atherosclerotic disease [59]. In addition, the results of these trials, namely DAPA-HF and EMPEROR-Reduced, have also introduced this new class of medications to the realm of cardiology and heart failure, as they do feature in the 2021 updated ACC HF treatment optimization report [60], though, as is often the case with novel classes of drugs, socioeconomic barriers and financial considerations may form impediments to broader uptake by patients.

The utilization of dual SGLT1/2 inhibitors may yield an added benefit due to the blockade of an additional variant of the sodium-glucose transporter; however, head-to-head trials comparing the two agents, perhaps with a third arm being assigned to GLP-1RA, may be needed in order to have enough data to structure the hierarchy of evidence regarding which agent would be ideal to start with (or indeed if combination therapy with both agents may yield added benefits compared to monotherapy).

INCRETINS

History of Incretins and their Mechanisms of Action

The insight into the role of incretins is derived from the observation that patients responded much better in terms of insulin secretion to orally ingested glucose than to intravenously administered glucose, which prompted the search for the endogenously produced compounds responsible for this phenomenon [61].

GLP1 is an insulinotropic compound produced and secreted from the intestine. It has a vital role in glucose metabolism as it moderates the glucose spike occurring after food intake by coordinating the insulin response and, thus, reducing uncontrolled spikes in glucose levels. This mechanism has been targeted by two essential classes of drugs: Dipeptidyl peptidase–4 inhibitors (DPP-4i) and GLP1-receptor agonists (GLP1-RA). GLP1-RA are analogs of the endogenously produced GLP1 molecules and are much more useful clinically due to their longer half-life, whereas DDP-4i aim at improving the half-life of the already-produced endogenous GLP-1 by preventing its break down. The first GLP-1RA, exenatide, was approved by the FDA in 2005, with the first DDP-4i, sitagliptin, following suit shortly thereafter in 2006 [61].

GLP1-RA mediate their glucoregulatory effects by stimulating the GLP-1 receptor, thus increasing insulin release after glucose intake. DPP-4i, on the other hand, prevents the breakdown and inactivation of both GLP-1 and glucose-dependent insulinotropic polypeptide, which increases the levels of these substances and likewise improves the insulin response to glucose loads. In addition to their glucoregulatory effect, GLP-1 has also been linked with endothelial function, myocardial metabolism, and ventricular contractility, which, coupled with glycemic control, may explain some of the beneficial cardiovascular effects they have in terms of cardiovascular outcomes [55]. Another essential advantage this class of drugs provides is a reduction in appetite and food intake, resulting in a loss of body weight (particularly in the case of GLP-1RA [62]), which is especially important in diabetes due to the role that obesity has to play in the disease process. Several trials were conducted to delineate the benefits of GLP-1RA and DPP-4i concerning glycemic control and cardiovascular outcomes, the most recent of which are discussed below.

Safety and Efficacy of DDP-4i

The SAVOR-TIMI-53 [63] trial, the largest of all trials evaluating DDP-4i drugs, was launched in 2010 and was designed to examine the cardiovascular effect of saxagliptin in patients with DM who were at high risk of cardiovascular events. A total of 16,492 patients were involved in the SAVOR trial, with a median follow-up of 2.1 years. The primary endpoints of the SAVOR-TIMI-53 trial were CVM, MI, or stroke, where it showed no evidence of benefit. Concerning glycemic control, HbA1c levels in the saxagliptin group were marginally lower across the trial (a reduction of 0.2%). Perhaps due to better glycemic control in the saxagliptin group, a reduction in albuminuria development/progression was achieved. However, patients in the saxagliptin group had significantly more hypoglycemic events and a sizeable increase in HF hospitalizations.

Another selective DPP-4i, alogliptin, was investigated in the EXAMINE [17] trial, which was conducted from 2009 to 2013. It sought to demonstrate the non-inferiority of alogliptin to placebo regarding cardiovascular events in patients with T2DM and a recent ACS. A total of 5,380 patients were recruited for this purpose, with a primary endpoint of CVM, MI, or stroke. The secondary safety endpoint included the components of the primary composite endpoint with the addition of urgent revascularization due to UA within 24 hours after hospital admission.

Reassuringly, the trial showed the non-inferiority of alogliptin to placebo for major cardiovascular events despite the high-risk patients recruited, though superiority was not demonstratable. In addition, alogliptin improved glycemic control as the HbA1c levels were reduced by 0.36%. The trial had a relatively modest duration of 1.5 years, which may have limited the observation of long-term benefits or risk of alogliptin and lessened its ability to reveal any cardiovascular benefits the drug may have, as this is the shortest time-duration of any of the DDP-4i clinical trials.

The TECOS [18] trial, which recruited over 14,500 patients with T2DM and established CVD, showed similar findings with 50 or 100 mg of sitagliptin, with non-inferiority being demonstrated in terms of a composite endpoint of CVM, MI, stroke or UA hospitalization. Furthermore, improved glycemic control was demonstrated in the intervention arm, as seen by an HbA1c reduction of 0.29%.

Another trial on DPP-4i drugs was that of Sheu *et al.*, which investigated the efficacy and optimal dosing regimen of MK-3102 [19] (omargliptin), a potent, selective, once-weekly DPP-4i. The trial included 685 patients who were randomly allocated to either placebo or omargliptin. The omargliptin group was subdivided into 5 dosing regimens (0.25 mg, 1 mg, 3 mg, 10 mg, 25 mg). The primary hypothesis to be tested was if treatment by omargliptin would provide a more significant reduction in HbA1c levels in comparison to placebo. The secondary endpoints studied were whether omargliptin provides a more significant reduction in 2-hour post-meal glucose and fasting plasma glucose in a dose-related manner after 12 weeks of treatment.

After 12 weeks, significant reductions in HbA1c levels were achieved and maintained across all dosing sub-groups of omargliptin, with the maximal decrease in HbA1c occurring between 18-46 weeks and a decline in efficacy in the following period. In addition, the secondary outcomes, post-meal glucose and fasting plasma glucose, showed a dose-related improvement in the omargliptin groups at 12-weeks.

The MARLINA study by Groop *et al.* was a relatively small study of 360 patients, which assessed the efficacy of 5 mg of linagliptin on renal outcomes in patients with T2DM and renal dysfunction (defined as an elevated urine albumin-t--creatinine ratio (UACR) or as albuminuria). Despite the usual improvement in HbA1c, the linagliptin arm showed no significant changes in their UACR or GFR in relation to the placebo arm.

The CARMELINA [21] trial aimed to assess the cardiovascular and renal safety of 5 mg of linagliptin in T2DM. There were 6,979 patients enrolled in this study with 2.2 years of follow-up, with the primary endpoint being CVM, MI, or stroke. Linagliptin was not inferior to placebo regarding the endpoint mentioned above; however, superiority could not be demonstrated. As in the previous trials, the linagliptin arm had a relatively greater significant reduction in glycated hemoglobin.

Another study involving linagliptin was CAROLINA [22], which examined the effect of 5 mg linagliptin *versus* glimepride (rather than placebo) on cardiovascular and glycemic outcomes. A total of 6,042 patients were randomly allocated into one of the two arms and subsequently followed up for 6.3 years. As in previous trials, the primary endpoint was time to first occurrence of CVM, MI, or stroke. The results were not significantly different from previous trials, with linagliptin showing non-inferiority to glimepride and a lack of superiority on further statistical probing. As the control group involved an active hypoglycemic agent, HbA1c reductions were roughly equal between the two groups; however, hypoglycemic events were significantly less likely in the linagliptin group (RRR of 77%; NNT of 4).

In the RENALIS study, Muskiet *et al.* tested the efficacy of 5 mg linagliptin *versus* 1 mg glimepride to reduce the degree of renal dysfunction in 46 patients with T2DM and a GFR of >60 ml/min/1.73m^2, with 23 being randomized to each arm of the trial. There were no significant HbA1c reductions between the two hypoglycemic agents. Moreover, no significant differences were noted in UACR or GFR. Although there may have been a signal for increased fractional sodium and potassium excretion indicating higher natriuresis, it did not reach statistical significance.

Trials Evaluating the Safety and Efficacy of GLP-1ras

The ELIXA [24] trial examined the safety and efficacy of lixisenatide (10 to 20 mg) on cardiovascular outcomes in patients with diabetes and recent ACS and was the first study conducted on this class of drugs powered to detect differences in hard cardiovascular outcomes, such as mortality, stroke or hospitalization. In this study, there were 6,068 patients with a median follow-up duration of 25 months, from 2013 to 2015. The primary endpoint was CVM, MI, stroke, or UA hospitalization, for which the non-inferiority criteria were met; however, superiority testing failed to show a statistically significant difference. Though the follow-up period mentioned above may not seem long enough to detect significant differences, the authors noted that the event-rate was sufficiently high enough. It is reasonable to conclude that no nonglycemic cardiovascular benefits could be had using lixisenatide. Lixisenatide also significantly improved glycemic control as measured by glycated hemoglobin.

Another vital trial in this realm, the LEADER [64] initiated in 2010, studied the effect of adding 1.8 mg of liraglutide on cardiovascular outcomes. There were 9,340 patients included in the trial, all of whom were patients with diabetes with CVD or risk factors thereof, and they were then divided into 2 groups, one received liraglutide at a dose of 1.8 mg and the other received placebo with a follow-up extending over 3.8 years.

The primary composite outcome was the CVM, MI, or stroke. There was a significant reduction of CVM in the liraglutide group (RRR of 22%; NNT of 77) as well as a reduction in ACM (RRR of 15%; NNT of 72). In addition, liraglutide lowered the rate of renal microvascular events. Improvements in weight and systolic blood pressure were also noted in the liraglutide group. Despite the impressive reductions in mortality, the composite endpoint of CVM, MI, or stroke did not show a statistically significant difference between the two groups. Regarding glycemic control, the reduction in HbA1c was accompanied by a significantly lower rate of severe hypoglycemia (RRR of 31%).

A similar study that assessed the cardiovascular safety and the long-term outcomes of semaglutide is SUSTAIN-6 [26], which was also conducted on relatively high-risk patients with T2DM. The trial's total patient population was 3,297 patients with a median observational time of 2.1 years. Participants were randomized to either semaglutide (maintenance dose of 0.5 mg or 1 mg achieved after a fixed-dose escalation procedure) or placebo.

The primary composite outcome was CVM, MI, or stroke, for which semaglutide showed an impressive benefit (RRR of 26%; NNT of 44). As for HbA1c levels, the semaglutide group achieved significant and sustained reduction compared to placebo with a similar hypoglycemia rate. Also, more patients assigned to placebo received anti-hyperglycemic agents and cardiovascular medications, including antihypertensives, diuretics, and lipid-lowering drugs, which may indicate an overall improved cardiovascular profile after taking semaglutide.

Focusing on microvascular outcomes, like the LEADER trial, the semaglutide group had a reduced risk of new/worsening nephropathy. However, semaglutide also significantly increased the risk of diabetic retinopathy, unlike in LEADER, where the differences did not reach statistical significance.

The FREEDOM-1 [27] trial investigated the efficacy and tolerability of 2 doses of ITCA 650 (exenatide at a dose of 40 or 60 mg administered as an automated pump) *versus* placebo in 460 patients who had T2DM for an average duration of 8.9 years and whose HbA1c levels were elevated despite consistent therapy for three or more months. ITCA 650 significantly improved HbA1c and fasting plasma glucose levels, in addition to a meaningful improvement in weight.

Regarding the differences between the efficacy of the two-dose of ITCA 650, HbA1c reduction was similar between the two doses. However, the higher dose showed higher overall efficacy in an aggregate outcome of glycemic control measures in addition to weight loss, without increased adverse effects. Mild significant improvements in the lipid parameters were also noted.

The HARMONY [28] trial was aimed to assess albiglutide's cardiovascular safety and efficacy (a GLP-1RA formed *via* the genetic fusion of modified human GLP-1 to human albumin) in patients with T2DM and CVD. There were 9,503 patients assigned randomly to either placebo or 30 to 50 mg of albiglutide who were followed up for a median of 1.5 years. The primary endpoint was a composite of CVM, MI, or stroke, which albiglutide managed to reduce significantly (RRR of 22%; NNT of 50). However, only myocardial infarction from the primary composite showed a statistically significant improvement in the albiglutide group, and, unlike the LEADER trial, it did not manage to show a statistically significant benefit when outcome analysis was restricted solely to the occurrence of CVM, which may have been partially attributable to a shorter follow-up period. Further, both systolic blood pressure and body weight improved in the albiglutide group. As in other trials, the authors noted a delay between initiation of treatment and the appearance of benefits, with the magnitude of improvement being proportional to the amount of time since initiation.

Requiring add-on glucose-lowering therapies was a more common finding in the placebo arm, which, coupled with the reduction in HbA1c bestowed by albiglutide, indicates an improved glycemic profile. Fortunately, unlike the SUSTAIN-6 trial, there was no evidence of new/worsening retinopathy in the albiglutide arm.

In the EXCSEL trial, which was aimed to assess the benefit of adding 2 mg of exenatide in a study recruiting 14,752 patients (16.2% of whom had HF at baseline) with T2DM and a history of CVD, exenatide showed great promise. Though the trial did not show an improvement in the primary outcome of CVM, MI, or stroke, it did show a significant benefit with regards to ACM (RRR of 14%; NNT of 100); however, due to the hierarchical scheme of the statistical analysis, this result was not deemed statistically significant.

Dulaglutide is a GLP-1RA comprised of two modified human GLP-1 molecules covalently linked to an IgG4 heavy chain. The REWIND [30] trial assessed whether the addition of 1.5 mg of dulaglutide to the medication regimen of middle-aged and older people with high-risk T2DM improves cardiovascular outcomes compared with placebo. A total of 12,133 patients were screened from August 2011 to August 2013, with follow-up ending in August 2018 and providing the most prolonged follow-up duration of all GLP-1RA trials.

The primary endpoint was a composite of CVM, MI, or stroke, dulaglutide significantly improved (RRR of 12%; NNT of 72), with a greater degree of benefit being derived from its beneficial effect on stroke over MI.

In 2019, Husain *et al.* published the results of the landmark (and aptly named) PIONEER-6 [31] trial. The trial enrolled nearly 3,200 high-risk T2DM patients over 2 years and allocated them to either semaglutide (target dose of 14 mg) or placebo. Similar to the EXCSEL trial, in which the primary composite endpoint was not statistically significant, PIONEER-6 failed to demonstrate a statistically significant difference in terms of superiority, though it did meet its target of non-inferiority. However, it also showed a significant improvement in ACM, this time with a considerably larger effect size (RRR of 49%; NNT of 72), in addition to reducing CVM by 51%. As in previous trials, these clinically oriented improvements were also accompanied by reduced HbA1c levels.

Summary and Synthesis of the Literature on Incretins

Collectively, GLP-1RA seems to exert significant cardiovascular benefits, at least in select patient populations, whereas DDP-4i, despite improved glycemic control, were generally unable to demonstrate a similar benefit for CVM, MI, or stroke outcomes. Furthermore, some DDP-4i agents, as with saxagliptin in the SAVOR-TIMI-53 trial, were associated with adverse cardiovascular outcomes such as an increase in HF hospitalizations.

Several previous meta-analyses had examined the association of DPP-4 inhibitors with risk for HF or HF hospitalization, with inconsistent findings [65 - 68]. However, according to a meta-analysis conducted in 2019 [69], there was no increase in HF hospitalizations with DPP-4i *versus* placebo. However, significant heterogeneity was associated with this finding, as can be observed from the discussion of the above trials, with some agents being more dangerous than others.

In conclusion, GLP1-RAs are among the first classes of anti-diabetes medications that improve glycemic control and demonstrate improved cardiovascular outcomes such as mortality and adverse vascular events. Furthermore, they are gaining prominence in the modern landscape of diabetes treatment, with recent guidelines recommending their addition to diabetes patients at risk of cardiovascular disease irrespective of glycemic control, as can be shown in Table **3** [70]. This paradigm shift has mainly been triggered by the realization that a medication's CV benefit may be independent of its HbA1c-reducing capacity. Therefore, treatment optimization for patients with diabetes is no longer solely aimed at keeping glycemic control measures within reasonable bounds, but, instead, a more holistic approach that aims to reduce the overall risk of CV adverse events is advocated. The key to this approach is the newer classes of medications that have demonstrated a significant risk reduction in large scale randomized trials and have therefore become advocated as being the standard of treatment.

DDP-4i drugs, on the other hand, have not yielded results that are robust in terms of cardiovascular endpoints. Some trials were associated with an increased rate of HF and there were reports of pancreatitis with their use. Further, due to their similar action mechanisms (with one class, GLP-1RAs, acting as direct agonists and the other class, DDP-4i, mimicking their action by prolonging the half-life of endogenous GLP-1), it is generally not recommended to combine them. Therefore, due to the impressive results seen in GLP-1RA trials, the DDP-4i class of drugs seems destined to have a relegated role in managing T2DM.

In addition to the above discussion of efficacy and adverse effects, it is also worth noting that there are some differences between the different drugs within each class in terms of practical utilization and structural formation. For instance, lixisenatide and exenatide are both exendin-4-based GLP-1RAs; however, lixisenatide is a short-acting GLP-1RA administered once daily, whereas the exenatide formulation used in the EXSCEL trial was a once-weekly formulation. These differences have significant practical implications for real-world employment, where matters of practicality and dosing schedules are often critical in determining drug adherence. In addition, due to exendin-4 being originally derived from lizard venom, both aforementioned drug compounds share a risk of triggering immune reactions, whereas the third drug in this class, such as liraglutide, which has more analogy to human GLP-1, is less likely to result in such reactions [71].

Table 3. Recommendations and guidelines of ACC/AHA and ESC/EAS for pharmacotherapeutic management of dyslipidemia in primary and secondary prevention.

Organization	Year/ Region	Recommendations for Primary Prevention	Recommendations for Secondary Prevention
ACC/AHA (1,2)	**2019 (Primary)& 2018 (Secondary)/ USA**	– In patients aged 20 to 75 with an LDL-C ≥ 190mg/dL, maximally tolerated statin therapy is indicated. – In adults aged 40 to 75 with diabetes, moderate-intensity statin therapy is indicated regardless of ASCVD 10-year risk score. In the presence of multiple ASCVD risk factors*, high-intensity statins are a reasonable option. – In adults with a ≥20% 10-year ASCVD risk-score, it is recommended to aim for an LDL-C reduction of ≥50%. – In adults at intermediate (between 7.5 and 20% 10-year ASCVD risk), moderate-intensity statins should be initiated in the context of a risk-discussion with the patient. LDL-C levels should be reduced by ≥30%; for optimal risk-reduction, a target of ≥50% may be preferrable.	– In patients ≤75 years with ASCVD, high-intensity statin therapy with an LDL-C reduction goal of ≥50% should be used. If intolerable, a moderate-intensity statin with a goal of 30 to 49% reduction in LDL-C is acceptable. If maximally-tolerated statin therapy does not lower LDL-C below 70 mg/dL, Ezetimibe may be added. – In very high-risk patients with ASCVD in whom maximally-tolerated statin therapy fails to lower LDL-C below 70 mg/dL,

(Table 3) cont.....

		Therapy-intensification is favored in the presence of risk-enhancing factors – In intermediate-risk adults in whom a CAC score obtained for purposes of risk-assessment is 0, postponement of statin therapy for 5 to 10 years and reassessment is a reasonable option in the absence of other risk-enhancing factors such as smoking, family history of premature heart disease or smoking. In cases of a CAC of 1 to 99 in a patient aged ≥55 years, or a CAC ≥100 (or ≥75th percentile), statins may be initiated. – In patients at borderline-risk (5 to 7.5%), moderate-intensity statins may be initiated in the presence of risk-enhancing factors.	Ezetimibe may be added. If LDL-C remains above 70 mg/dL, a PCSK9 inhibitor may be initiated following a safety and cost-benefit discussion. – Both moderate and high intensity statin therapy are reasonable for adults aged ≥75 depending on an evaluation for potential benefit, adverse effects and patient preferences. – In patients with HFrEF due to ischemic disease who are not already on an ASCVD-indicated statin, moderate statins may be initiated.
ESC/EAS (3)	**2019/ Europe**	– In patients at very high risk, a reduction of LDL-C by ≥50% and to <55mg/dL should be aimed for. Initiation of high-intensity LDL---lowering regimens or intensification of existing ones is likely to be required. – In high risk patients, an LDL-C reduction of ≥50% and to a target of <70mg/dL should be aimed for. – In moderate risk individuals, aim for an LDL-C <100mg/dL – In low risk individuals, aim for an LDL-C <116mg/dL. – A high-intensity statin should be used (up to the maximally-tolerated dose) in order to achieve the aforementioned targets. – If statin therapy fails to achieve these targets, Ezetemibe should be added. – If dual-therapy with statins and Ezetimibe is not able to achieve targets, and the patient is very high risk, PCSK9 addition may be considered.	– In patients with established ASCVD, a reduction of LDL-C by ≥50% and to <55mg/dL should be aimed for. Initiation of high-intensity LDL-C-lowering regimens or intensification of existing ones is likely to be required. – If statin therapy fails to achieve these targets, Ezetemibe should be added. – If dual-therapy with statins and Ezetimibe is not able to achieve targets, and the patient is at very-high risk, PCSK9 addition is recommended.

Abbreviations: ACC, American College of Cardiology; AHA, American Heart Association; ESC, European Society of Cardiology; EAS, European Atherosclerosis Society; ASCVD, Atherosclerotic Cardiovascular Disease; PCSK9, Proprotein convertase subtilisin/kexin type 9; HFrEF, Heart Failure with reduced Ejection Fraction.

INSULIN

A Historical Overview of the Discovery of Insulin

Few discoveries in medicine have been as revolutionary as that of insulin. First administered to Leonard Thompson in 1922, it became commercially available the next year, and thereafter, a number of innovations improving its pharmacokinetic profile have made it a bedrock of diabetes treatment in those with T1DM and certain cases of T2DM.

Few discoveries in medicine have been as revolutionary as that of insulin. Frederick Banting, a returning officer from WW1, had recently joined the academia at the University of Toronto after a somewhat unsuccessful experience at his private clinic, where patient visits were relatively scarce. Along with his student, Charles Best, they extracted insulin from the canine pancreas, and by January 11, 1922, the first doctor to have the privilege of administering insulin, Ed Jeffrey, used it on a 14-year-old by the name of Leonard Thompson. Though the initial injection was not as successful as one would have hoped, a subsequent injection series on January 23, 1922 resulted in marked clinical improvement, and the first instance of insulin use on humans was forever written down in the annals of history [72, 73].

Thereafter, the course of insulin development rapidly accelerated. Just one year later, by 1923, insulin was commercially available in the US after a partnership between Eli Lilly and Company and Banting's team. By 1936, the crystallization of insulin allowed the production of protamine zinc insulin, a product with a longer half-life that revolutionized insulin administration due to a pharmacokinetic profile allowing longer-lasting effects, which made night-time dosing significantly more convenient [73]. In the following decades, the formulation of longer-acting forms of insulin, culminating in what is known as "basal insulins," has allowed for improved glycemic control in patients. Ever since these events, insulin has had a life-saving role in T1DM and, in some cases, is needed for those with T2DM.

The Role of Glycemic Control In Outcome-optimization and a Discussion of Trials on Recent Insulin Formulations

Several studies [74 - 77], including the landmark UKDPS glycemic control trial [77], have suggested that administering early basal insulin at levels sufficient to normalize the plasma glucose levels may safely improve outcomes in diabetic patients. However, it is worth noting that the trial mentioned above did not show a statistically significant benefit in macrovascular outcomes or deaths related to diabetes, with most of the evidence deriving from microvascular improvements

[78]. In addition, the evidence in this area has not been free of conflicts or heterogeneity, with the ACCORD trial showing an increase in mortality with more intensive regimens [79].

The Outcome Reduction with an Initial Glargine Intervention (ORIGIN) [32] trial tested the previous hypothesis. The ORIGIN trial involved 12,573 patients older than 50 years of age with cardiovascular risk factors and early T2DM or impaired glucose tolerance and was conducted over 8 years. The trial primarily aimed to uncover whether insulin administration would improve outcomes in these patients, as well as reduce the risk of incidents of T2DM in those with impaired glucose tolerance. The cardiovascular outcomes of this trial were divided into two primary composites. The first was CVM, MI, and stroke. The second was a revascularization procedure or hospitalization for heart failure.

The trial showed no significant differences in either outcome between the two groups. However, basal insulin glargine in the ORIGIN trial reduced the new onset of diabetes despite causing an increase in weight. Although some concern may be raised that the reduction in incident diabetes was merely an illusory one due to hyperglycemia masking by glargine, the authors remarked that this was quite unlikely considering the mean duration of action is about one day. Unfortunately, the reduction in new-onset diabetes did not come free of cost, as the incidence of severe hypoglycemic events considerably increased in the intervention arm.

Another trial, DEVOTE 7 [33], tested the cardiovascular safety and the risk of severe hypoglycemia with degludec (2nd-generation basal insulin) *versus* glargine U100 (a 1st-generation basal insulin) in over 7,600 patients with T2DM and a high risk of cardiovascular events. The primary outcome was the occurrence of CVM, MI, or stroke. The trial's secondary outcome was severe hypoglycemia as defined by the American Diabetes Association guidelines: an episode requiring the infusion of carbohydrates, glucagon, or other corrective actions.

Results from DEVOTE 7 showed that although the risk of MACE and death in patients with T2DM who were at high risk of CVD were comparable, degludec had the benefit of significantly reducing the risk of hypoglycemia, a result probably attributable to its pharmacodynamic properties and relatively long half-life. In addition, though HbA1c reductions were not significantly different, the degludec group had lower fasting glucose levels.

The CONCLUDE trial, which tested two 2nd generation basal insulins, namely degludec *versus* 300 U/mL of glargine, in patients with T2DM and an increased risk of hypoglycemia, showed that the incidence of nocturnal symptomatic and severe hypoglycemic events was significantly reduced with degludec use;

however, the overall rates of symptomatic hypoglycemia did not differ at a statistically significant level between the two groups [80].

In the ACHIEVE trial, Menegheni *et al.* compared 300 U/mL of glargine with two older 1st generation basal insulins (glargine 100 U/mL and detemir, termed "standard of care"), and showed that although, as expected, glycemic control as measured by HbA1c levels did not differ significantly, a greater proportion of patients in the glargine 300 U/mL arm were able to achieve their HbA1c levels without episodes of clinically significant hypoglycemia [35].

Hypoglycemia is a significant source of fear and anxiety for both patients and physicians alike, and therefore this fear not only decreases the functional capacity of patients with diabetes, but may also act as a barrier to improving glycemic control; thus, reducing its incidence is of vital importance [81]. Furthermore, there has been some debate about whether or not it may be linked with adverse cardiovascular outcomes, with an analysis of the Veterans Affairs Diabetes Trial (VADT) [82] showing a sizeable association between recent severe hypoglycemic attacks and impending CVM, with an HR of 3.72. Nevertheless, this observation has been somewhat contradicted by other data pieces, and it is not entirely clear whether hypoglycemia is a pathophysiological cause for adverse cardiovascular events or merely a high-risk marker [82]. Nevertheless, severe hypoglycemia, and its associated adverse effects, highlight the importance of striking a balance between not allowing glucose levels to exceed acceptable levels while at the same time not being so stringent with the targeted goals to the point of inadvertently increasing the harming the patient.

A more recent trial is that of ORMD-0801 [36], a novel insulin injection trialed on a total of 354 patients with T2DM with an HbA1c above 7.5% who did not manage to achieve glycemic control despite the use of multiple medications. Though the full paper is not yet available at the time of writing, it seems that this newer insulin formulation may have improved HbA1c levels without significant differences in the incidence of hypoglycemic events, although we hesitate to draw any conclusions in the absence of the full paper.

In the ICODEC [37] trial, recently published in 2020, Rosenstock *et al.* demonstrated the similar efficacy of a once-weekly insulin injection and daily glargine 100 U/mL in terms of glycemic control in 247 patients with T2DM who had not previously received insulin therapy, with no statistically significant difference between the two groups with respect to HbA1c reduction. Notably, the Icodec group had a higher incidence of hypoglycemic alerts, though no statistically significant difference in severe/clinically-significant hypoglycemic events was observed. This trial's revolutionary potential is that it promises to significantly reduce the number of insulin injections patients with diabetes may have to use and, thus, significantly improves patient's compliance to treatment.

Overall, newer formulations of insulin have done much to control the incidence of severe hypoglycemic attacks and to ensure greater stability of glucose levels. The use of once-weekly injections as with Icodec is another step that may also make insulin therapy much more palatable and convenient to patients. Nevertheless, a limitation is that the ICODEC trial only compared this new insulin formulation to an older generation basal insulin (glargine 100 U) instead of the more recent generation of basal insulins (glargine 300 U and degludec). Therefore, it is difficult to judge whether this more convenient dosing schedule may come at the cost of a higher incidence of hypoglycemic attacks, given the decreased hypoglycemic risk of the latter class of drugs or not.

ALPHA-GLUCOSIDASE INHIBITORS

Alpha-glucosidase inhibitors work by inhibiting the enzymes present in the intestine's brush border that are responsible for the breakdown of disaccharides and polysaccharides. Since the intestine is only capable of absorbing monosaccharides, these enzymes' inhibition decreases post-prandial glucose levels and is a theoretical benefit in cardiovascular outcomes [83].

In the ACE [38] study, Holman *et al.* studied the effect of 150 mg of acarbose in over 6,500 patients with CHD and impaired glucose tolerance. There was no statistically significant difference in the primary outcome of CVM, MI, stroke, ACS, surgical/endovascular intervention, or amputation, and the improvement in glycated hemoglobin was marginal (0.06%). However, there was a benefit in that acarbose reduced the incidence of new-onset diabetes (RRR of 18%). In modern clinical practice, however, the gastrointestinal adverse effects of flatulence and diarrhea, in addition to the availability of other therapeutic options, have limited their clinical uptake and they are not very widely prescribed.

THIAZOLIDINEDIONES

Thiazolidinediones (TZD) entered the US in 1996, with troglitazone being the first to gain FDA approval. However, their history has been a rocky one, as the agent mentioned above, the first to be clinically utilized, was withdrawn only four years later due to hepatotoxicity [84]. The two TZDs still in clinical use today are pioglitazone and rosiglitazone. TZDs act as agonists at peroxisome proliferator-activated receptors (PPAR), improving insulin sensitivity, and blood pressure and reducing coagulation factors [85 - 87]. However, TZDs are not without downsides and are known to cause fluid retention, HF, and, especially in rosiglitazone, higher LDL concentrations [86 - 88].

To assess pioglitazone's adequacy in improving clinically-oriented hard endpoints, including cardiovascular morbidity and/or mortality, conduction of the PROactive study took place [89]. The PROactive trial included 5,238 patients randomly assigned to either pioglitazone (15 to 45 mg) or placebo. The trial's primary endpoints were as follows: ACM, MI, stroke, ACS, or endovascular or surgical intervention. Simultaneously, the secondary endpoints included ACM, MI (excluding silent MI), and stroke.

Pioglitazone did not show significant evidence of benefit for the primary composite endpoint. However, the secondary endpoint was reduced significantly (RRR of 16%; NNT of 49). Furthermore, better glycemic control, HDL concentration, triglyceride, and blood pressure levels were noted in the pioglitazone group; however, a slight increase in LDL was also observed.

Another trial, this time aimed at uncovering the clinical utility of rosiglitazone, was the RECORD [39] trial, which was done to prove rosiglitazone's non-inferiority in combination with either metformin or sulfonylurea compared with metformin and dual sulfonylurea therapy for cardiovascular outcomes. It recruited 7,428 patients and had a primary outcome of time to the first occurrence of cardiovascular hospitalization or CVM.

The RECORD trial results suggested that CVM and CV hospitalizations were not significantly different between the two groups; however, there was a significant increase in a composite of HF-related death or hospitalization, possibly due to the increased fluid retention causing an increase in cardiac afterload. Indeed, participants in the rosiglitazone group were more likely to be prescribed loop diuretics, which may have been related to HF incidence. Serious adverse events due to hyperglycemia were less frequent with rosiglitazone usage, but hypoglycemic severe adverse events and bone fractures were more likely to occur.

Overall, the adverse cardiovascular outcomes reported in the RECORD trial have resulted in pioglitazone taking the front-seat as the TZD of choice in many settings where TZD therapy is to be initiated; however, the use of TZDs in general has been in decline, with both rosiglitazone and pioglitazone seeing drops in their rates of prescription [90]. These findings, combined with the rise of more recent hypoglycemic agents with a better profile of cardiovascular benefits, which we will soon touch upon, make it unlikely that TZDs will have a prominent role in the management of diabetes going forward.

Comparison of the Different Novel Modalities and Critical Gaps in the Literature

With the recent influx of modern anti-diabetic drugs, several of which have had demonstrable cardiovascular benefits, an important question arises from this newfound luxury: which of these drugs is of greater efficacy? The ideal study design to answer these questions is, of course, a randomized clinical trial; however, to our knowledge, no such trials have been published thus far.

An alternative is to leverage the widespread use of these drugs to conduct real-life comparisons of patient outcomes when they are assigned to the different types of drugs. Obviously, the observational nature of such studies limits their reliability due to a number of potential biases. In order to minimize the risk of confounding factors muddling the actual outcomes caused directly by drug activity, a recent analytic approach that has been in the ascendancy is propensity-matching.

In brief, this method aims to identify a sizeable number of patients taking each type of drug, identify the various factors that influence their prognosis, and select pairs of highly similar patients in terms of these essential prognostic factors. Although this method is far from perfect, it is invaluable to glean some insight into answers that no clinical trials conducted so far have provided. It should also be noted that such methods allow us to leverage real-life data from a more representative sample of patients than do clinical trials, as the latter often recruit patients who do not always exactly correspond to the types of patients one might encounter in the clinic. In addition, large-scale collaborations between different groups with facilitated network meta-analyses incorporating individual patient data may also allow for a more informed decision derived from the higher-quality evidence found in randomized control trials (RCTs), though possible heterogeneity between the settings and patient populations of different trials would have to be considered in such an analysis.

One such study is EMPRISE [91], which used real-world databases to compare the efficacy of empagliflozin, sitagliptin, and GLP-1RA. In their analysis, Patorno *et al.* showed that empagliflozin could reduce HF hospitalizations compared to the latter two. Concerning ACM, empagliflozin had demonstrably better outcomes than sitagliptin, with an HR of 0.52. Compared with GLP-1RA, the HR of 0.79, though favoring empagliflozin, had too wide of a confidence interval (0.6 to 1.05) to establish statistical significance.

For now, it seems that SGLT2i may have the edge over GLP-1RA and that both classes seem far more advantageous in terms of clinical outcomes than DDP-4i; however, further clinical trials are needed to cement these findings. In addition, as several patients with diabetes require multi-drug therapy, it would be enlightening to establish whether the combination of SGLT2i and GLP-1RA could have a synergistic/additive effect in terms of improving patient outcomes. Indeed, Clegg *et al.* analyzed the synergistic/additive effect of both once-weekly GLP-1RA and SGLT2i using propensity score matching compared to placebo or patients only receiving once-weekly GLP-1 RA exenatide to estimate the relative efficacy of a GLP1 receptor agonist (exenatide weekly) *versus* that of an SGLT2i (primarily dapagliflozin) using data from the EXSCEL trial [92]. They demonstrated the likely benefit on cardiovascular outcomes and mortality with lower risk of MACE compared to controls [92]. Furthermore, there is a need for data comparing traditional SGLT2i and dual SGLT2i in order to determine which of the two classes is more effective. Once such questions have been answered, it may be necessary to investigate further the effects of each specific drug falling under each class, as our previous discussion has shown that there has been some degree of heterogeneity between the different trials conducted on agents within the same class of drugs. Such questions highlight the advances that we have made in the treatment of diabetes, as not too long ago, we did not even possess sufficient data regarding the benefit of many anti-glycemic agents in improving cardiovascular outcomes; whereas now, the focus seems to be on finding out which of these drugs is most efficacious.

LIPID-LOWERING THERAPIES STATINS

Historical Context and Foundational Trials (Pre-2010)

Over the decades, our understanding—and subsequently treatment—of dyslipidemia and atherosclerotic disease has made leaps and bounds, and there has perhaps been a no finer example of these advances than the discovery and subsequent incorporation of statins into modern clinical practice.

The first statin compound, mevastatin, discovered by Akira Endo *et al.* in 1976, was a fungal antibiotic compound inhibiting cholesterol synthesis [93]. Later, Brown *et al.*'s Nobel-winning discovery uncovered the mechanism by which statins do this: they inhibit HMG-CoA reductase, thus prevent the conversion of HMG-CoA into mevalonate, a critical step in the biosynthesis pathway of cholesterol [94, 95]. In 1987, the first medically utilized statin, lovastatin, was granted FDA approval seven years after the publication of a paper by Alet al; reporting on its inhibitory effects on cholesterol synthesis [96].

Nevertheless, it was not until ten years later that statin use started cementing a foothold in clinical practice with the results of the Scandinavian Simvastatin Survival Study in 1994 that showed an improvement in survival (RRR of 30%; NNT of 25) with the use of 20 mg of simvastatin in patients with a prior MI and angina (in addition to a cholesterol level between 212 and 309 mg/dL) [97].

The next secondary prevention landmark clinical trial was PROVE-IT TIMI [98], published in 2004. The trial showed that using atorvastatin 80 mg, as opposed to 40 mg of pravastatin, significantly reduced ACM's composite outcome, stroke, revascularization, MI, or UA (16% RRR; NNT of 26). Other vital clinical trials demonstrating benefit in secondary prevention include the TNT trial [99], showing the superiority of 80 mg *versus* 10 mg of atorvastatin in patients with CHD (RRR of 31%; NNT of 31) and the SPARCL trial [100], recruiting those with a history of stroke or transient ischemic attacks (stroke recurrence RRR of 15%; NNT of 53).

Notable primary prevention trials include AFCAPS/TexCAPS [101], which recruited those without a history of CHD but an LDL of above 150 mg/dL and showed a significant lowering of the occurrence of a composite of MI, UA, or sudden cardiac death (RRR of 37%; NNT of 24). Another critical trial in the realm of primary prevention was CARDS [102], which showed that in people with diabetes with no CHD and with an LDL of below 160 mg/dL, 10 mg of atorvastatin resulted in an impressive reduction in the composite of CVM, MI, stroke and revascularization (RRR of 37%; NNT of 31). The Heart Protection Study [103] recruited patients with total cholesterol above 135 mg/dL who had some additional risk factors (ASCVD, diabetes, or hypertension). The trial results were positive, with a decrease in ACM (RRR of 13%; NNT of 77) and major vascular events (RRR of 24%; NNT of 19).

The ASCOT-LLA trial [104], which recruited patients who had hypertension, an LDL <251 mg/dL, and three of several risk factors for CVD, reported a sizeable reduction using 10 mg of atorvastatin in the outcome of CVM/MI (RRR of 36%; NNT of 91), though the absolute difference in events was not as large as the previously discussed trials, as observed by the NNT of 91. A sub-analysis conducted on just over 2,500 hypertensive diabetic patients with no history of CHD played a key role in extending statin therapy to those who have diabetes [105].

A controversial trial in the domain of primary prevention, JUPITER [106], attempted to expand the use of statins not just to those with elevated LDL and/or other traditional risk factors, but to those with higher levels of the inflammatory biomarker C-reactive protein (CRP), on the premise that due to decisive role of inflammation in atherosclerosis, those with elevated CRP levels would be at a higher risk of ASCVD and thus more likely to benefit from statin therapy. The trial recruited those whose CRP level was 2 mg/dL and above while excluding those whose LDL was below 130 mg/dL. The investigators used 20 mg of rosuvastatin, and the reported results were an impressive reduction in the composite outcome of CVM, MI, cerebrovascular events, revascularization, or

UA hospitalization (RRR of 44%; NNT of 83). Nevertheless, there were several critiques concerning the trial's conduct regarding its premature termination and potential conflict of interests.

To avoid painting too biased a picture and outline the limitations of statin therapy, we must also mention trials that did not significantly benefit. These include the 4D trial [107], conducted on diabetic patients receiving hemodialysis, in whom 20 mg of atorvastatin showed no appreciable benefit on a composite outcome of CVM, MI, or stroke; likewise, the AURORA trial [108], conducted on patients receiving hemodialysis in general (rather than solely people with diabetes), showed no significant improvement of the same endpoint using 10 mg of rosuvastatin.

Likewise, the limitations of statin therapy were apparent in the CORONA [109] and GISSI-HF [110] trials, which similarly showed no improvements in ACM using 10 mg of rosuvastatin in patients with heart failure; however, it is worth noting that the former of these trials showed a reduction in the total hospitalizations (due to any cause) and a reduction in the number of patients requiring hospitalization for cardiovascular causes (HR of 0.92; NNT of 48).

Contemporary Analyses and Trials (2010 and Onwards)

Beyond these foundational trials, of which we only mentioned some, a more recent meta-analysis of 26 studies, including nearly 170,000 patients, published in 2010 by the cholesterol treatment trialists (CTT) allows us greater insights into the efficacy of statins [111]. The results showed a 10% decrease in ACM for every 39 mg/dL (equating to one mmol/l) reduction in LDL-C, mainly driven by the decrease in deaths due to CHD (RR of 20%), with no significant benefit seen in deaths due to stroke. In a primary-prevention analysis of those with an ASCVD score of between 5 and 10%, a similar 11% decrease in ACM with a more considerable decrease in vascular events. In a subsequent meta-analysis published in 2012, the investigators focused on patients at a lower risk for vascular disease [112]. In the lowest risk group of patients, those with an ASCVD of below 5%, there was an impressive 38% RRR in major vascular events but no significant reduction in overall or vascular deaths. The meta-analyses also revealed several limitations: first, the benefit in those older than 75 years was somewhat lower (RRR of 14%), and the reduction in LDL achieved by doubling the dose of statins was only 6%, highlighting the difficulty in achieving LDL targets in patients whose levels are significantly elevated by merely increasing the dose of statins.

Contemporarily, several trials have furthered our understanding of statin-based therapy and the limitations thereof. The more recent SHARP trial [113], published in 2011, showed a significant decrease in the primary composite outcome of

CVM, MI, stroke, or revascularization (RRR of 17%; NNT of 48) in patients with chronic kidney disease. The trial was not sufficiently powered to investigate these outcomes in the subset of patients on dialysis, with only 33% of patients being on maintenance dialysis; therefore, it did not manage to settle the question of statin utility in dialysis patients or provide enough evidence to overturn the evidence from the aforementioned AURORA and 4D trials. A more recent 2016 meta-analysis by the CTT analyzing data from just over 7,000 dialysis patients showed no significant benefit of starting or intensifying statin therapy in dialysis patients [114].

Another outstanding question not answered by the aforementioned landmark trials was that of the role of ezetimibe in addition to statin therapy and whether it had any role in further improving outcomes beyond using statins alone. To that end, the IMPROVE-IT trial [115], published in 2015, recruited over 18,000 patients hospitalized for recent ACS within ten days whose LDL levels were above 50 mg/dL (with an upper limit of 100 or 125 mg/dL depending on whether or not they were maintained on lipid-lowering therapy, respectively) and randomized them to 40 mg of simvastatin alone or in combination with 10 mg of ezetimibe. The results showed a reduction in the occurrence of a primary endpoint composed of CVM, MI, stroke, UA, or revascularization (RRR of 6.4%; NNT of 50).

Regarding the question of statin utility in elderly patients, a recent 2017 post-hoc analysis of the ALLHAT trial analyzed the benefit of 40 mg of simvastatin for primary prevention in adults aged 65 years and older [116]. Interestingly, the results verged on evidence for harm for the overall cohort (HR of 1.18 and a 95% CI of 0.97 to 1.42) in those aged 75 and above in terms of ACM (HR of 1.34 and a 95% CI of 0.98 to 1.84); however, the result is based post-hoc subgroup analysis obviously limits the conclusions one can draw.

Overall, the abundance of clinical data since the inception of statins has allowed them to rise to the peak of lipid-lowering drugs' hierarchy, becoming the most widely used such class. Accordingly, recent guidelines, of which a summary can be found in Table **2**, rely on statins as the mainstay of treatment in both primary and secondary prevention, with other drugs being relegated to an ancillary role when further LDL lowering is desired.

PCSK9 INHIBITORS

Rationale and Initial Discoveries

The initial impetus for the development of Proprotein convertase subtilisin/Kexin type 9 inhibitors (PCSK9i) was derived from observational data regarding the role of PCSK9 variants on LDL levels and subsequent vascular disease. In 2003, a paper published in Nature-Genetics by a team of researchers at the University of Paris reported a form of familial hypercholesterolemia caused by a gain-o--function mutation in the gene PCSK9 [117]. Subsequently, Cohen *et al.* reported in 2006 that individuals with nonsense mutations (*i.e.* mutations causing a cessation of protein translation) had 28% lower LDL levels and a massive 88% RRR for the outcome of CHD [118].

To further elucidate the mechanism by which PCSK9 inhibition may be beneficial, Grefhorst *et al.*'s experiments reported in 2008 showed that mice injected with recombinant PCSK9 had markedly increased levels of LDL-receptor degradation and subsequently expressed far fewer LDL-receptors on liver's surface [119]; therefore, PCSK9 inhibition would conversely result in an upregulation of LDL-receptors on the liver and increased clearance rates of these particles from the plasma, thus lowering LDL levels and subsequent atherosclerotic events.

Trials Demonstrating LDL-Lowering Efficacy

To that end, a number of trials have investigated the ability of PCSK9i to lower LDL levels in a variety of settings. GAUSS 1 and 2, which were 12-week RCTs involving statin-intolerant patients and published in 2012 and 2014, respectively, provided important insight into the usefulness of PCSK9i in statin-intolerant hypercholesterolemic patients, who present a clinical challenge as they are unable to receive the most important lipid-lowering class of drugs. The results were positive, with impressive reductions in LDL levels across different doses of evolocumab and irrespective of the use of ezetimibe (See Table **2**) [40, 42]. The findings of the original GAUSS-1 study, which was a phase II study recruiting patients intolerant to only one type of statin, were further bolstered by the GAUSS-2, which was a phase III study recruiting only those intolerant to at least two types of statins.

The RUTHERFORD trial, published in 2014, reported similarly impressive results over 12 weeks in those suffering from familial hypercholesterolemia using 140 mg or 420 mg of evolocumab, with LDL reductions of 59.2% and 61.3%, respectively [41]. In contrast to the GAUSS studies, which were smaller studies with only 12 weeks of follow up, the larger DESCARTES trial, a phase III study recruiting 901 patients and conducted over 52-weeks, provided more robust, long-term follow-up data over a larger number of patients with similarly significant lowering of LDL-levels as seen previously [43]. MENDEL-2, a 12-week phase III study that recruited 564 patients with hypercholesterolemia and a Framingham risk score of ≤10%, also demonstrated a large degree of LDL lowering compared to both placebo and ezetimibe [44].

In LAPLACE-2, Robinson *et al.* focused on a clinically challenging strata of patients: do PCSK9i have a role in patients whose LDL-targets were not met with statins? Moreover, do they provide more promise than ezetimibe in that regard? Participants were included regardless of whether the failure of LDL-lowering therapy was attributable to being completely statin-intolerant or whether the current dose could not achieve target levels. In the largest trial up to that point with nearly 1900 patients, evolocumab was significantly more capable of reducing LDL than both placebo and ezetimibe. Regardless of the maximally tolerated intensity of the statins that patients were on, the analysis was stratified according to the type of background statin therapy [45].

YUKAWA, conducted on high-risk Japanese patients treated with statins regardless of ezetimibe usage (Refer to Table **2** for the exact recruitment criteria), was a relatively small study of 307 patients that similarly showed significant LDL reductions compared to placebo [46].

Evidence of Clinical Efficacy and Reduction of Hard Outcomes

Subsequently, PCSK9i were approved in 2015, a mere three years following the first trial (GAUSS-1) publication, demonstrating their efficacy at lowering LDL. Nevertheless, a clinically important question persisted: just how effective are PCSK9i in improving clinically-oriented patient outcomes? There was useful data that strongly linked LDL-lowering to clinical outcomes, as briefly discussed in the meta-analysis by the CTT mentioned above; however, physicians were nevertheless eagerly awaiting results showing whether PCSK9i's impressive LDL-lowering abilities translated into improved clinical endpoints.

To fill this gap in the literature, Sabatine *et al.* conducted FOURIER [48], a massive RCT conducted over 168-weeks, recruiting over 27,000 patients whose LDL levels were above 70 mg/dL despite statin therapy. The results, published in 2017, were an impressive reduction in two clinically important endpoints: firstly, it reduced the composite outcome of CVM, MI, stroke, hospitalization for UA, and coronary revascularization (HR of 0.85; NNT of 40). A more challenging secondary end-point consisting of only the first three aforementioned outcomes was likewise improved (HR of 0.8; NNT of 67). As in the previous trials, the addition of evolocumab significantly lowered LDL levels.

Two years later, Szarek *et al.* published their results in the New England Journal of Medicine, showing similarly impressive results with 75 mg of Alirocumab [49]. The trial recruited nearly 19,000 patients and showed significant reductions in CVM (HR of 0.83) and non-fatal stroke, MI, UA hospitalization, and revascularization (HR of 0.87).

In contrast to evolocumab and alirocumab, which are both wholly humanized monoclonal antibodies, bococizumab is only partially humanized. Therefore, the recipient's immune system is more prone to recognizing it as a foreign antigen and thus produce neutralizing antibodies that thus reduce its efficacy. This reasoning may explain the results of the SPIRE trial, published by Ridker *et al.* in 2017. The SPIRE trial consisted of two parts: SPIRE-1, in which patients had a minimum baseline LDL of 70 mg/dL, and SPIRE-2, in which patients had a minimum baseline of 100 mg/dL.

Unsurprisingly, the event rate was higher in the latter trial. In addition, it also had a longer follow-up time (12 as opposed to 7 months). These factors may have combined to produce a sharp contrast between the two trials: SPIRE-2 resulted in a significant reduction in the primary endpoint with an HR of 0.79, whereas SPIRE-1 did not.

Ultimately, the trial was prematurely canceled due to the sponsoring company deciding to no longer pursue the development of bococizumab, a decision mainly attributable to the significant fraction of patients (29% as per the discussion section of the paper) developing neutralizing antibodies.

Despite their impressive LDL-lowering capacities, a paramount concern when discussing the utility of any class of drugs—especially novel ones—is cost-efficacy. PCSK9i, owing to their novelty and relatively rapid approval, face a challenge familiar to a variety of other novel therapeutics across a number of fields in medicine. Their price, particularly in resource-poor settings, halts their widespread adoption relative to cheaper supplementary LDL-lowering therapeutics such as ezetimibe. Indeed, in the 2018 secondary prevention AHA/ACC guidelines, it was remarked that PCSK9i offered relatively low value-for-benefit in light of their price. However, reductions in the price of evolocumab prompted an analysis by Fonarow *et al.* They argued the cost-effectiveness of evolocumab following the 2018 price reductions [120].

In conclusion, PCSK9i has added a vital tool to the armament of physicians who can employ it in patients in whom LDL-control is unsatisfactory with statins (due to either insufficient magnitude at the highest intensities or an inability to utilize higher intensities as a result of intolerance). Furthermore, the recent trials demonstrating improvements in clinically significant endpoints mentioned above further bolster our trust in these drugs' clinical efficacy and their ability to improve patient-oriented outcomes. A summary of the essential PCSK9i trials can be found in Table **2**.

OMEGA-3 FATTY ACIDS

Much of the interest in the role of omega-3 fatty acids initially arose from observational data linking increased fish consumption rates to lower cardiovascular events [121]. The first major trial investigating this question was GISSI-P, which recruited more than 11,000 patients with a recent MI and randomized them to a placebo or 1 g of EPA/DHA [122]. The trial demonstrated some benefit in this group of patients; however, its lack of blinding raised some concern regarding the validity of its conclusions. In addition, as the standard of care in treating ischemic heart disease had improved and the baseline event rate decreased, there was skepticism regarding the potential benefits of additional therapies. The next sizeable randomized trial to investigate was the OMEGA trial [122], published in 2010 and recruiting 3,851 patients. Unlike the GISSI-P trial, the results were disappointing as there was no reduction in statistically significant reduction in ACM, SCD, cerebrovascular or cardiovascular events, or revascularization procedures; however, this may, at least partially, be attributable to the underpowered nature of the study.

ORIGIN [51], published in 2012 and recruiting a population of over 12,000 patients with diabetes or impaired glucose tolerance at high risk for CVD (patients must have had a previous history of significant ASCVD or at high risk thereof), was similarly negative in that there was no statistically significant benefit in ACM, CVM, arrhythmias or vascular events, despite a modest triglyceride reduction of 14.5 mg/dL in the group randomized to 1 g of omega-3 fatty acids.

The more recent ASCEND [52] trial had also investigated the benefit of 1 g of omega-3 supplements in people with diabetes and similarly found little evidence of statistically significant benefit in any ACM, revascularization, or severe vascular events in a trial including over 15,000 patients and spanning six years. Unlike ORIGIN, which did allow patients with a previous history of CVD into the trial, ASCEND only allowed people with diabetes with no history of CVD.

Despite these recent disappointments, the REDUCE-IT trial results published in 2019 did breathe new life into the use of omega-3 supplements in that it showed significant clinical benefit in a large trial spanning seven years and including over 8,000 patients [53]. However, there were several key differences in both the intervention and the inclusion criteria, which merit discussion: first, the dose used was 4 g, which is much higher than the 1 g used in ASCEND and ORIGIN, and certainly much higher than the doses commonly used daily by laypeople. Second, all participants were required to have elevated triglyceride levels, a requirement absent in either of the two previously mentioned trials. Third, the intervention arm only included a highly purified ester form of eicosapaentanoic acid (EPA), with no added docoshaexanoic acid (DHA). In addition, the population recruited was a relatively high-risk one as patients were required to be above 45 with a history of established CVD or patients with diabetes above 50 with at least one additional risk factor. In terms of lipid levels, there was a reduction in both triglyceride levels (19.7%) and an amelioration of the increase in LDL levels (6.6%). Clinically speaking, this translated into impressive benefits in terms of a composite outcome of CVM, MI, stroke, revascularization, or UA (HR of 0.75; NNT of 21) and a similarly impressive reduction in an even more stringent clinical endpoint of CVM, MI or stroke (HR of 0.74; NNT of 28).

Nevertheless, another anticipation in the studies of omega-3 fatty acids awaited as the most recent trial results at the time of writing, STRENGTH [54], were published in 2020. The trial recruited just over 13,000 patients over a span of 6 years that recruited patients with previous CVD for secondary prevention or high-risk primary prevention patients (Refer to Table **2** for the exact criteria). The trial, like REDUCE-IT, employed a 4 g dose of omega-3 acids; however, unlike REDUCE-IT, it also included DHA within the formulation. The results were similar in terms of triglyceride reduction, with an 18.1% additional decrease being observed in the intervention arm, however, the trial failed to demonstrate any significant benefit in the composite endpoint of CVM, MI, stroke, revascularization, or UA hospitalization. The trial was eventually halted due to futility as clinical evidence of benefit seemed exceedingly unlikely; in addition, there was evidence of a slightly higher risk of atrial fibrillation in the intervention arm.

Some authors have attributed the sharp contrast in the trials to the placebos used in each trial: REDUCE-IT employed mineral oil as a comparator, whereas STRENGTH employed corn oil. In their discussion of the STRENGTH trial, Nicholls *et al.* mention that mineral oil is not an entirely neutral comparator. It may adversely affect lipid profiles and thus give Icosapent Ethyl (the formulation employed in REDUCE-IT) an apparent benefit due to being a harmful placebo. The authors support this conclusion by stating that there was no evidence of such an adverse effect on lipid profile in the control arm of the STRENGTH trial, although the authors do admit that it may be difficult to attribute all of Icosapent Ethyl's apparent benefits to the adverse effect of mineral oil used as placebo. In addition, a subsequent study by Lakshmanan *et al.* seemed to show that plaque volumes did not seem to correlate with mineral oil placebo intake [123]. A recent meta-analysis on the subject of mineral oil also seems to support the conclusion that it is doubtful for the large effect sizes observed in REDUCE-IT to be attributable to mineral oil [124].

Another noticeable difference that may help explain the divergence in outcomes is DHA's addition in STRENGTH; however, the authors state that this is unlikely considering that its plasma concentrations increased by only moderate amounts and did not strongly correlate with event rates.

Overall, there is conflicting evidence regarding the efficacy of omega-3 fatty acids, with the heterogeneity between comparators and the different formulations used in trials limiting one's ability to draw definite conclusions regarding its clinical benefit; nevertheless, it may well have an important role to play in high-risk patients with hypertriglyceridemia, although future trials may be needed to clarify the exact benefits/downsides of EPA and DHA and what the ideal omega-3 formulation is. A summary of the most important recent clinical trials can be found in Table **2**.

CONCLUSION

For antidiabetic medications, Over the past decade, the advent of newer therapies, namely SGLT2 inhibitors and GLP-1 agonists, has revolutionized the treatment of type 2 diabetes over the past decade. These medications not only improve glycemic indices, as previous treatments did, but also significantly decrease the risk of cardiovascular events.

Nevertheless, important challenges remain unaddressed. First, these medications have not yet been compared in head-to-head trials, and it is therefore unclear which may be optimal to start with. Second, these medications have been studied mainly in the setting of type 2 diabetes, as such, their benefit in type 1 diabetes remains unclear. Finally, the expense of these drugs relative to older treatments will likely delay their adoption in resource-limited settings. Despite these challenges, these drugs have opened a new frontier in the treatment of diabetes and will provide new hope for millions of patients worldwide.

Moreover, in lipid-lowering medications, the use of statins in the treatment of dyslipidemia represented a historical milestone in the prevention of cardiovascular disease. Because of their affordability and demonstrable clinical benefit, they are likely to remain the cornerstone of treatment for the foreseeable future. The arrival of newer treatments, such as Omega-3 fatty acids and PCSK9 inhibitors, are important advances that complement the use of statins.

The former may represent an important option in patients with hypertriglyceridemia, although controversies surrounding the conflicting results of the STRENGTH and REDUCE-IT trials remain to abound. The latter is important for patients in whom statins are insufficient and/or intolerable, although their expense remains a barrier to greater adoption. The judicious use of these treatments, alone or in combination, will protect millions globally from a prevalent and potentially fatal condition.

CONSENT FOR PUBLICATION

Not applicable.

CONFLICT OF INTEREST

The authors declare no conflict of interest, financial or otherwise.

ACKNOWLEDGEMENTS

Declared none.

REFERENCES

[1] Fatima S, Jameel A, Ayesha F, Menzies DJ. The shifting paradigm in the treatment of type 2 diabetes mellitus-A cardiologist's perspective. Clin Cardiol 2017; 40(11): 970-3. [http://dx.doi.org/10.1002/clc.22781] [PMID: 28841228]

[2] Gubitosi-Klug RA, Lachin JM, Backlund JYC, Lorenzi GM, Brillon DJ, Orchard TJ. Intensive diabetes treatment and cardiovascular outcomes in type1 diabetes: The DCCT/EDIC study 30-year follow-up. Diabetes Care 2016; 39(5): 686-93. [http://dx.doi.org/10.2337/dc15-1990] [PMID: 26861924]

[3] Livingstone SJ, Looker HC, Hothersall EJ, Wild SH, Lindsay RS, Chalmers J, *et al.* 2012.https://dx.plos.org/10.1371/journal.pmed.1001321

[4] Constantino MI, Molyneaux L, Limacher-Gisler F, *et al.* Long-term complications and mortality in young-onset diabetes: type 2 diabetes is more hazardous and lethal than type 1 diabetes. Diabetes Care 2013; 36(12): 3863-9.http://care.diabetesjournals.org/cgi/doi/10.2337/dc12-2455 [http://dx.doi.org/10.2337/dc12-2455] [PMID: 23846814]

[5] Tancredi M, Rosengren A, Svensson A-M, *et al.* Excess Mortality among Persons with Type 2 Diabetes. N Engl J Med 2015; 373(18): 1720-32.http://www.nejm.org/doi/10.1056/NEJMoa1504347 [http://dx.doi.org/10.1056/NEJMoa1504347] [PMID: 26510021]

[6] Schnell O, Rydén L, Standl E, Ceriello A. Updates on cardiovascular outcome trials in diabetes. Cardiovasc Diabetol 2017; 16(1): 128. [http://dx.doi.org/10.1186/s12933-017-0610-y] [PMID: 29020969]

[7] Zinman B, Wanner C, Lachin JM, *et al.* Empagliflozin, Cardiovascular Outcomes, and Mortality in Type 2 Diabetes. N Engl J Med 2015; 373(22): 2117-28.http://www.nejm.org/doi/10.1056/NEJMoa1504720 [http://dx.doi.org/10.1056/NEJMoa1504720] [PMID: 26378978]

[8] Neal B, Perkovic V, Mahaffey KW, *et al.* Canagliflozin and Cardiovascular and Renal Events in Type 2 Diabetes. N Engl J Med 2017; 377(7): 644-57.http://www.nejm.org/doi/10.1056/NEJMoa1611925 [http://dx.doi.org/10.1056/NEJMoa1611925] [PMID: 28605608]

[9] Perkovic V, Jardine MJ, Neal B, *et al.* Canagliflozin and Renal Outcomes in Type 2 Diabetes and Nephropathy. N Engl J Med 2019; 380(24): 2295-306. [http://dx.doi.org/10.1056/NEJMoa1811744] [PMID: 30990260]

[10] Wiviott SD, Raz I, Bonaca MP, *et al.* Dapagliflozin and Cardiovascular Outcomes in Type 2 Diabetes. N Engl J Med 2019; 380(4): 347-57.

[http://dx.doi.org/10.1056/NEJMoa1812389] [PMID: 30415602]

[11] McMurray JJV, Solomon SD, Inzucchi SE, *et al.* Dapagliflozin in Patients with Heart Failure and Reduced Ejection Fraction. N Engl J Med 2019; 381(21): 1995-2008. [http://dx.doi.org/10.1056/NEJMoa1911303] [PMID: 31535829]

[12] Cosentino F, Cannon CP, Cherney DZI, *et al.* Efficacy of Ertugliflozin on Heart Failure-Related Events in Patients With Type 2 Diabetes Mellitus and Established Atherosclerotic Cardiovascular Disease: Results of the VERTIS CV Trial. Circulation 2020; 142(23): 2205-15. [http://dx.doi.org/10.1161/CIRCULATIONAHA.120.050255] [PMID: 33026243]

[13] Packer M, Anker SD, Butler J, *et al.* Cardiovascular and Renal Outcomes with Empagliflozin in Heart Failure. N Engl J Med 2020; 383(15): 1413-24. [http://dx.doi.org/10.1056/NEJMoa2022190] [PMID: 32865377]

[14] Bhatt DL, Szarek M, Steg PG, Cannon CP, Leiter LA, McGuire DK, *et al.* Sotagliflozin in Patients with Diabetes and Recent Worsening Heart Failure. N Engl J Med 2020; 1-11. [PMID: 33200892]

[15] Bhatt DL, Szarek M, Pitt B, Cannon CP, Leiter LA, McGuire DK, *et al.* Sotagliflozin in Patients with Diabetes and Chronic Kidney Disease. N Engl J Med 2020; 1-11. [PMID: 33200891]

[16] Scirica BM, Bhatt DL, Braunwald E, *et al.* Saxagliptin and cardiovascular outcomes in patients with type 2 diabetes mellitus. N Engl J Med 2013; 369(14): 1317-26. [http://dx.doi.org/10.1056/NEJMoa1307684] [PMID: 23992601]

[17] White WB, Cannon CP, Heller SR, Nissen SE, Bergenstal RM, Bakris GL, *et al.* Alogliptin after acute coronary syndrome in patients with type 2 diabetes. Austrian J Clin Endocrinol Metab 2014; 7(2): 77.

[18] Green JB, Bethel MA, Armstrong PW, *et al.* Effect of Sitagliptin on Cardiovascular Outcomes in Type 2 Diabetes. N Engl J Med 2015; 373(3): 232-42. [http://dx.doi.org/10.1056/NEJMoa1501352] [PMID: 26052984]

[19] Sheu WHH, Gantz I, Chen M, *et al.* Safety and efficacy of omarigliptin (MK-3102), a novel Once-Weekly DPP-4 Inhibitor for the treatment of patients with type 2 diabetes. Diabetes Care 2015; 38(11): 2106-14. [http://dx.doi.org/10.2337/dc15-0109] [PMID: 26310692]

[20] Groop PH, Cooper ME, Perkovic V, *et al.* Linagliptin and its effects on hyperglycaemia and albuminuria in patients with type 2 diabetes and renal dysfunction: the randomized MARLINA-T2D trial. Diabetes Obes Metab 2017; 19(11): 1610-9. [http://dx.doi.org/10.1111/dom.13041] [PMID: 28636754]

[21] Rosenstock J, Perkovic V, Johansen OE, *et al.* Effect of Linagliptin vs Placebo on Major Cardiovascular Events in Adults With Type 2 Diabetes and High Cardiovascular and Renal Risk: The CARMELINA Randomized Clinical Trial. JAMA 2019; 321(1): 69-79. [http://dx.doi.org/10.1001/jama.2018.18269] [PMID: 30418475]

[22] Rosenstock J, Kahn SE, Johansen OE, *et al.* Effect of Linagliptin vs Glimepiride on Major Adverse Cardiovascular Outcomes in Patients With Type 2 Diabetes: The CAROLINA Randomized Clinical Trial. JAMA 2019; 322(12): 1155-66. [http://dx.doi.org/10.1001/jama.2019.13772] [PMID: 31536101]

[23] Muskiet MHA, Tonneijck L, Smits MM, *et al.* Effects of DPP-4 inhibitor linagliptin *versus* sulfonylurea glimepiride as add-on to metformin on renal physiology in overweight patients with type 2 diabetes (Renalis): A randomized, double-blind trial. Diabetes Care 2020; 43(11): 2889-93. [http://dx.doi.org/10.2337/dc20-0902] [PMID: 32900785]

[24] Pfeffer MA, Claggett B, Diaz R, *et al.* Lixisenatide in Patients with Type 2 Diabetes and Acute Coronary Syndrome. N Engl J Med 2015; 373(23): 2247-57. [http://dx.doi.org/10.1056/NEJMoa1509225] [PMID: 26630143]

[25] Marso SP, Daniels GH, Brown-Frandsen K, *et al.* Liraglutide and Cardiovascular Outcomes in Type 2 Diabetes. N Engl J Med 2016; 375(4): 311-22. [http://dx.doi.org/10.1056/NEJMoa1603827] [PMID: 27295427]

[26] Marso SP, Bain SC, Consoli A, *et al.* Semaglutide and Cardiovascular Outcomes in Patients with Type 2 Diabetes. N Engl J Med 2016; 375(19): 1834-44. [http://dx.doi.org/10.1056/NEJMoa1607141] [PMID: 27633186]

[27] Rosenstock J, Buse JB, Azeem R, *et al.* Efficacy and safety of ITCA 650, a novel drug-Device GLP-1 receptor agonist, in type 2 diabetes uncontrolled with oral antidiabetes drugs: The FREEDOM-1 trial. Diabetes Care 2018; 41(2): 333-40. [http://dx.doi.org/10.2337/dc17-1306] [PMID: 29242349]

[28] Hernandez AF, Green JB, Janmohamed S, *et al.* Albiglutide and cardiovascular outcomes in patients with type 2 diabetes and cardiovascular disease (Harmony Outcomes): a double-blind, randomised placebo-controlled trial. Lancet 2018; 392(10157): 1519-29. [PMID: 30291013]

[29] Holman RR, Bethel MA, Mentz RJ, *et al.* Effects of Once-Weekly Exenatide on Cardiovascular Outcomes in Type 2 Diabetes. N Engl J Med 2017; 377(13): 1228-39. [http://dx.doi.org/10.1056/NEJMoa1612917] [PMID: 28910237]

[30] Gerstein HC, Colhoun HM, Dagenais GR, *et al.* Dulaglutide and cardiovascular outcomes in type 2 diabetes (REWIND): a double-blind, randomised placebo-controlled trial. Lancet 2019; 394(10193): 121-30. [http://dx.doi.org/10.1016/S0140-6736(19)31149-3] [PMID: 31189511]

[31] Husain M, Birkenfeld AL, Donsmark M, *et al.* Oral Semaglutide and Cardiovascular Outcomes in Patients with Type 2 Diabetes. N Engl J Med 2019; 381(9): 841-51. [http://dx.doi.org/10.1056/NEJMoa1901118] [PMID: 31185157]

[32] Gerstein HC, Bosch J, Dagenais GR, *et al.* Basal insulin and cardiovascular and other outcomes in dysglycemia. N Engl J Med 2012; 367(4): 319-28. [http://dx.doi.org/10.1056/NEJMoa1203858] [PMID: 22686416]

[33] Marso SP, McGuire DK, Zinman B, *et al.* Efficacy and Safety of Degludec *versus* Glargine in Type 2 Diabetes. N Engl J Med 2017; 377(8): 723-32. [http://dx.doi.org/10.1056/NEJMoa1615692] [PMID: 28605603]

[34] Philis-Tsimikas A, Stratton I, Nørgård Troelsen L, Anker Bak B, Leiter LA. Efficacy and Safety of Degludec Compared to Glargine 300 Units/mL in Insulin-Experienced Patients With Type 2 Diabetes: Trial Protocol Amendment (NCT03078478). J Diabetes Sci Technol 2019; 13(3): 498-506.http://journals.sagepub.com/doi/10.1177/1932296819841585 [http://dx.doi.org/10.1177/1932296819841585] [PMID: 30974986]

[35] Meneghini L, Blonde L, Gill J, *et al.* Insulin glargine 300 U/mL *versus* first-generation basal insulin analogues in insulin-naïve adults with type 2 diabetes: 12-month outcomes of ACHIEVE Control, a prospective, randomized, pragmatic real-life clinical trial. Diabetes Obes Metab 2020; 22(11): 1995-2003. [http://dx.doi.org/10.1111/dom.14116] [PMID: 32538550]

[36] Eldor Roy, Fleming GA, Neutel J, Homer KE, Kidron M, Rosenstock J. 105-LB: Evening Oral Insulin (ORMD-0801) Glycemic Effects in Uncontrolled T2DM Patients. Diabetes 2020; 69 (Supplement 1): 105. http://diabetes.diabetesjournals.org/content/69/Supplement_1/105-LB

[37] Rosenstock J, Bajaj HS, Janež A, *et al.* Once-Weekly Insulin for Type 2 Diabetes without Previous Insulin Treatment. N Engl J Med 2020; 383(22): 2107-16. [http://dx.doi.org/10.1056/NEJMoa2022474] [PMID: 32960514]

[38] Holman RR, Coleman RL, Chan JCN, *et al.* Effects of acarbose on cardiovascular and diabetes outcomes in patients with coronary heart disease and impaired glucose tolerance (ACE): a randomised,

double-blind, placebo-controlled trial. Lancet Diabetes Endocrinol 2017; 5(11): 877-86. [PMID: 28917545]

[39] Home PD, Pocock SJ, Beck-Nielsen H, *et al.* Rosiglitazone evaluated for cardiovascular outcomes in oral agent combination therapy for type 2 diabetes (RECORD): a multicentre, randomised, open-label trial. Lancet 2009; 373(9681): 2125-35. [http://dx.doi.org/10.1016/S0140-6736(09)60953-3] [PMID: 19501900]

[40] Sullivan D, Olsson AG, Scott R, *et al.* Effect of a monoclonal antibody to PCSK9 on low-density lipoprotein cholesterol levels in statin-intolerant patients: the GAUSS randomized trial. JAMA 2012; 308(23): 2497-506. [http://dx.doi.org/10.1001/jama.2012.25790] [PMID: 23128163]

[41] Raal FJ, Stein EA, Dufour R, *et al.* PCSK9 inhibition with evolocumab (AMG 145) in heterozygous familial hypercholesterolaemia (RUTHERFORD-2): a randomised, double-blind, placebo-controlled trial. Lancet 2015; 385(9965): 331-40. [http://dx.doi.org/10.1016/S0140-6736(14)61399-4] [PMID: 25282519]

[42] Stroes E, Colquhoun D, Sullivan D, *et al.* Anti-PCSK9 antibody effectively lowers cholesterol in patients with statin intolerance: the GAUSS-2 randomized, placebo-controlled phase 3 clinical trial of evolocumab. J Am Coll Cardiol 2014; 63(23): 2541-8. [http://dx.doi.org/10.1016/j.jacc.2014.03.019] [PMID: 24694531]

[43] Blom DJ, Hala T, Bolognese M, *et al.* A 52-week placebo-controlled trial of evolocumab in hyperlipidemia. N Engl J Med 2014; 370(19): 1809-19. [http://dx.doi.org/10.1056/NEJMoa1316222] [PMID: 24678979]

[44] Koren MJ, Lundqvist P, Bolognese M, *et al.* Anti-PCSK9 monotherapy for hypercholesterolemia: the MENDEL-2 randomized, controlled phase III clinical trial of evolocumab. J Am Coll Cardiol 2014; 63(23): 2531-40. [http://dx.doi.org/10.1016/j.jacc.2014.03.018] [PMID: 24691094]

[45] Robinson JG, Nedergaard BS, Rogers WJ, *et al.* Effect of evolocumab or ezetimibe added to moderate- or high-intensity statin therapy on LDL-C lowering in patients with hypercholesterolemia: the LAPLACE-2 randomized clinical trial. JAMA 2014; 311(18): 1870-82. [http://dx.doi.org/10.1001/jama.2014.4030] [PMID: 24825642]

[46] Hirayama A, Honarpour N, Yoshida M, *et al.* Effects of evolocumab (AMG 145), a monoclonal antibody to PCSK9, in hypercholesterolemic, statin-treated Japanese patients at high cardiovascular risk--primary results from the phase 2 YUKAWA study. Circ J 2014; 78(5): 1073-82. [http://dx.doi.org/10.1253/circj.CJ-14-0130] [PMID: 24662398]

[47] Ridker PM, Revkin J, Amarenco P, *et al.* Cardiovascular Efficacy and Safety of Bococizumab in High-Risk Patients. N Engl J Med 2017; 376(16): 1527-39. [http://dx.doi.org/10.1056/NEJMoa1701488] [PMID: 28304242]

[48] Sabatine MS, Giugliano RP, Keech AC, *et al.* Evolocumab and Clinical Outcomes in Patients with Cardiovascular Disease. N Engl J Med 2017; 376(18): 1713-22. [http://dx.doi.org/10.1056/NEJMoa1615664] [PMID: 28304224]

[49] Szarek M, White HD, Schwartz GG, *et al.* Alirocumab Reduces Total Nonfatal Cardiovascular and Fatal Events: The ODYSSEY OUTCOMES Trial. J Am Coll Cardiol 2019; 73(4): 387-96. [http://dx.doi.org/10.1016/j.jacc.2018.10.039] [PMID: 30428396]

[50] Koren MJ, Sabatine MS, Giugliano RP, *et al.* Long-Term Efficacy and Safety of Evolocumab in Patients With Hypercholesterolemia. J Am Coll Cardiol 2019; 74(17): 2132-46. [http://dx.doi.org/10.1016/j.jacc.2019.08.1024] [PMID: 31648705]

[51] Bosch J, Gerstein HC, Dagenais GR, *et al.* n-3 fatty acids and cardiovascular outcomes in patients with dysglycemia. N Engl J Med 2012; 367(4): 309-18. [http://dx.doi.org/10.1056/NEJMoa1203859] [PMID: 22686415]

[52] Bowman L, Mafham M, Wallendszus K, *et al.* Effects of n-3 Fatty Acid Supplements in Diabetes Mellitus. N Engl J Med 2018; 379(16): 1540-50. [http://dx.doi.org/10.1056/NEJMoa1804989] [PMID: 30146932]

[53] Bhatt DL, Steg PG, Miller M, *et al.* Cardiovascular Risk Reduction with Icosapent Ethyl for Hypertriglyceridemia. N Engl J Med 2019; 380(1): 11-22. [http://dx.doi.org/10.1056/NEJMoa1812792] [PMID: 30415628]

[54] Nicholls SJ, Lincoff AM, Garcia M, *et al.* Effect of High-Dose Omega-3 Fatty Acids *vs* Corn Oil on Major Adverse Cardiovascular Events in Patients at High Cardiovascular Risk: The STRENGTH Randomized Clinical Trial. JAMA -. JAMA 2020; 44195(22): 1-13. [http://dx.doi.org/10.1001/jama.2020.22258]

[55] Ni L, Yuan C, Chen G, Zhang C, Wu X, Zhang C, *et al.* SGLT2i: beyond the glucose-lowering effect. Cardiovasc Diabetol 2020; 19(1): 98. [http://dx.doi.org/10.1186/s12933-020-01071-y] [PMID: 32590982]

[56] Ehrenkranz JRL, Lewis NG, Kahn CR, Roth J. Phlorizin: a review. Diabetes Metab Res Rev 2005; 21(1): 31-8. [http://dx.doi.org/10.1002/dmrr.532] [PMID: 15624123]

[57] Cefalo CMA, Cinti F, Moffa S, *et al.* Sotagliflozin, the first dual SGLT inhibitor: current outlook and perspectives. Cardiovasc Diabetol 2019; 18(1): 20. [http://dx.doi.org/10.1186/s12933-019-0828-y] [PMID: 30819210]

[58] Bhatt DL, Szarek M, Pitt B, *et al.* Sotagliflozin in Patients with Diabetes and Chronic Kidney Disease. N Engl J Med :Internet. 2014/02/06. 2021 Nov 18;384(2):129–39. Available from: https://www.ahajournals.org/doi/10.1161/CIR.0000000000000625

[59] Das SR, Everett BM, Birtcher KK, *et al.* 2018 ACC Expert Consensus Decision Pathway on Novel Therapies for Cardiovascular Risk Reduction in Patients With Type 2 Diabetes and Atherosclerotic Cardiovascular Disease: A Report of the American College of Cardiology Task Force on Expert Consensus Decision Pathways. J Am Coll Cardiol 2018; 72(24): 3200-23. [http://dx.doi.org/10.1016/j.jacc.2018.09.020] [PMID: 30497881]

[60] Yancy CW, Januzzi JL Jr, Allen LA, *et al.* 2017 ACC Expert Consensus Decision Pathway for Optimization of Heart Failure Treatment: Answers to 10 Pivotal Issues About Heart Failure With Reduced Ejection Fraction: A Report of the American College of Cardiology Task Force on Expert Consensus Decision Pathways. J Am Coll Cardiol 2018; 71(2): 201-30. [http://dx.doi.org/10.1016/j.jacc.2017.11.025] [PMID: 29277252]

[61] White JR. A Brief History of the Development of Diabetes Medications. Diabetes :Spectr [Internet]. 2014 May;27(2):82 LP-- 86. Available from: http://spectrum.diabetesjournals.org/content/ 27/2/82.abstract

[62] Brunton S. 2014.https://pubmed.ncbi.nlm.nih.gov/24499291

[63] Špinar J, Smahelová A. SAVORTIMI 53 -Saxagliptin and cardiovascular outcomes in patients with type 2 diabetes mellitus. Vnitr Lek 2013; 59(11): 1003-7. [PMID: 24279445]

[64] Mann JFE, Nauck MA, Nissen SE, Pocock S, Ph D, Zinman B, *et al.* Liraglutide and cardiovascular outcomes in type 2 diabetes. Drug Ther Bull 2016; 54(9): 101.

[65] Verma S, Goldenberg RM, Bhatt DL, Farkouh ME, Quan A, Teoh H, *et al.* 2017. http://cmajopen.ca/lookup/doi/10.9778/cmajo.20160058

[66] Li L, Li S, Deng K, *et al.* Dipeptidyl peptidase-4 inhibitors and risk of heart failure in type 2 diabetes: systematic review and meta-analysis of randomised and observational studies. BMJ 2016; 352: i610.https://www.bmj.com/lookup/doi/10.1136/bmj.i610 [http://dx.doi.org/10.1136/bmj.i610] [PMID: 26888822]

[67] Patil HR, Al Badarin FJ, Al Shami HA, *et al.* Meta-analysis of effect of dipeptidyl peptidase-4

inhibitors on cardiovascular risk in type 2 diabetes mellitus. Am J Cardiol 2012; 110(6): 826-33.https://linkinghub.elsevier.com/retrieve/pii/S0002914912013446 [http://dx.doi.org/10.1016/j.amjcard.2012.04.061] [PMID: 22703861]

[68] Monami M, Ahrén B, Dicembrini I, Mannucci E. Dipeptidyl peptidase-4 inhibitors and cardiovascular risk: a meta-analysis of randomized clinical trials. Diabetes Obes Metab 2013; 15(2): 112-20.http://doi.wiley.com/10.1111/dom.12000 [http://dx.doi.org/10.1111/dom.12000] [PMID: 22925682]

[69] Alfayez OM, Almutairi AR, Aldosari A, Al Yami MS. Update on Cardiovascular Safety of Incretin-Based Therapy in Adults With Type 2 Diabetes Mellitus: A Meta-Analysis of Cardiovascular Outcome Trials. Can J Diabetes 2019; 43(7): 538-545.e2. [http://dx.doi.org/10.1016/j.jcjd.2019.04.003] [PMID: 31175007]

[70] 9. Pharmacologic Approaches to Glycemic Treatment: *Standards of Medical Care in Diabetes-2020.* Diabetes Care 2020; 43 (Suppl. 1): S98-S110.http://care.diabetesjournals.org/lookup/doi/10.2337/dc20-S009 [http://dx.doi.org/10.2337/dc20-S009] [PMID: 31862752]

[71] Shamriz O, NaserEddin A, Mosenzon O, Sviri S, Tal Y. Allergic reaction to exenatide and lixisenatide but not to liraglutide: Unveiling anaphylaxis to glucagon-like peptide 1 receptor agonists. Diabetes Care 2019; 42(9): e141-2. [http://dx.doi.org/10.2337/dc19-0720] [PMID: 31296642]

[72] Vecchio I, Tornali C, Bragazzi NL, Martini M. The discovery of insulin: An important milestone in the history of medicine. Front Endocrinol (Lausanne) 2018; 9(October): 613. [http://dx.doi.org/10.3389/fendo.2018.00613] [PMID: 30405529]

[73] White JR. A Brief History of the Development of Diabetes Medications. Diabetes Spectr 2005; 27(2): 82-6. Available from: http://spectrum.diabetesjournals.org/content/27/2/82.abstract.

[74] Zoungas S, Arima H, Gerstein HC, *et al.* Effects of intensive glucose control on microvascular outcomes in patients with type 2 diabetes: a meta-analysis of individual participant data from randomised controlled trials. Lancet Diabetes Endocrinol 2017; 5(6): 431-7. [http://dx.doi.org/10.1016/S2213-8587(17)30104-3] [PMID: 28365411]

[75] Reaven PD, Emanuele NV, Wiitala WL, *et al.* Intensive Glucose Control in Patients with Type 2 Diabetes - 15-Year Follow-up. N Engl J Med 2019; 380(23): 2215-24. [http://dx.doi.org/10.1056/NEJMoa1806802] [PMID: 31167051]

[76] Kirkman MS, Mahmud H, Korytkowski MT. Intensive Blood Glucose Control and Vascular Outcomes in Patients with Type 2 Diabetes Mellitus. Endocrinol Metab Clin North Am 2018; 47(1): 81-96. [http://dx.doi.org/10.1016/j.ecl.2017.10.002] [PMID: 29407058]

[77] Intensive Blood-Glucose Control with Sulfonylureas or Insulin Compared with Conventional Treatment and Risk of Complications in Patients with Type 2 Diabetes. Endocrinologist 1999; 9(2): 149. [http://dx.doi.org/10.1097/00019616-199903000-00016]

[78] Intensive blood-glucose control with sulphonylureas or insulin compared with conventional treatment and risk of complications in patients with type 2 diabetes (UKPDS 33). Lancet 1998; 352(9131): 837-53. [http://dx.doi.org/10.1016/S0140-6736(98)07019-6] [PMID: 9742976]

[79] Gerstein HC, Miller ME, Byington RP, *et al.* Effects of intensive glucose lowering in type 2 diabetes. N Engl J Med 2008; 358(24): 2545-59. [http://dx.doi.org/10.1056/NEJMoa0802743] [PMID: 18539917]

[80] Philis-Tsimikas A, Klonoff DC, Khunti K, *et al.* Risk of hypoglycaemia with insulin degludec *versus* insulin glargine U300 in insulin-treated patients with type 2 diabetes: the randomised, head-to-head CONCLUDE trial. Diabetologia 2020; 63(4): 698-710.

[http://dx.doi.org/10.1007/s00125-019-05080-9] [PMID: 31984443]

[81] Abdelfattah OM, Hassanein M, Saad AM, Abela G, Aldasouqi S. Fasting- Evoked En route Hypoglycemia in Diabetes (FEEHD): From guidelines to clinical practice. Curr Diabetes Rev 2020; 16(9): 949-56.http://www.eurekaselect.com/178134/article
[http://dx.doi.org/10.2174/1573399816666200107103829] [PMID: 31914915]

[82] Frier BM, Schernthaner G, Heller SR. Hypoglycemia and cardiovascular risks. Diabetes Care 2011; 34 (SUPPL. 2).
[http://dx.doi.org/10.2337/dc11-s220]

[83] Hedrington MS, Davis SN. Considerations when using alpha-glucosidase inhibitors in the treatment of type 2 diabetes. Expert Opin Pharmacother 2019; 20(18): 2229-35.
[http://dx.doi.org/10.1080/14656566.2019.1672660] [PMID: 31593486]

[84] Krishnaswami A, Ravi-Kumar S, Lewis JM. Thiazolidinediones: a 2010 perspective. Perm J 2010; 14(3): 64-72.https://pubmed.ncbi.nlm.nih.gov/20844707
[http://dx.doi.org/10.7812/TPP/09-052] [PMID: 20844707]

[85] Vasudevan AR, Balasubramanyam A. Thiazolidinediones: a review of their mechanisms of insulin sensitization, therapeutic potential, clinical efficacy, and tolerability. Diabetes Technol Ther 2004; 6(6): 850-63.
[http://dx.doi.org/10.1089/dia.2004.6.850] [PMID: 15684639]

[86] Peters Harmel AL, Kendall DM, Buse JB, Boyle PJ, Marchetti A, Lau H. Impact of adjunctive thiazolidinedione therapy on blood lipid levels and glycemic control in patients with type 2 diabetes. Curr Med Res Opin 2004; 20(2): 215-23.
[http://dx.doi.org/10.1185/030079903125002937] [PMID: 15006017]

[87] Khan S, Khan S, Imran M, Pillai KK, Akhtar M, Najmi AK. Effects of pioglitazone and vildagliptin on coagulation cascade in diabetes mellitus--targeting thrombogenesis. Expert Opin Ther Targets 2013; 17(6): 627-39.
[PMID: 23356568]

[88] Nesto RW, Bell D, Bonow RO, *et al.* Thiazolidinedione use, fluid retention, and congestive heart failure: a consensus statement from the American Heart Association and American Diabetes Association. October 7, 2003. Circulation 2003; 108(23): 2941-8.
[http://dx.doi.org/10.1161/01.CIR.0000103683.99399.7E] [PMID: 14662691]

[89] Dormandy JA, Charbonnel B, Eckland DJ, *et al.* Secondary prevention of macrovascular events in patients with type 2 diabetes in the PROactive Study (PROspective pioglitAzone Clinical Trial In macroVascular Events): a randomised controlled trial. Lancet 2005; 366(9493): 1279-89.https://linkinghub.elsevier.com/retrieve/pii/S0140673605675289
[http://dx.doi.org/10.1016/S0140-6736(05)67528-9] [PMID: 16214598]

[90] Hashim S, Gomes T, Juurlink D, Hellings C, Mamdani M. The rise and fall of the thiazolidinediones: impact of clinical evidence publication and formulary change on the prescription incidence of thiazolidinediones. J Popul Ther Clin Pharmacol = J la Ther des Popul la Pharmacol Clin 2013; 20(3): e238-42.

[91] Elisabetta P, Ajinkya P, Jessica F, Mehdi N, Deborah W, Anouk D-L, *et al.* Abstract 11928: Cardiovascular Effectiveness of Empagliflozin Compared to DPP4 Inhibitors and to GLP1 Receptor Agonists: Interim Analysis From the Empagliflozin Comparative Effectiveness and Safety (EMPRISE) Study. Circulation 2019; 140 (Suppl. 1): A11928-8. [Internet].
[http://dx.doi.org/10.1161/circ.140.suppl_1.11928]

[92] Clegg LE, Penland RC, Bachina S, *et al.* Effects of exenatide and open-label SGLT2 inhibitor treatment, given in parallel or sequentially, on mortality and cardiovascular and renal outcomes in type 2 diabetes: insights from the EXSCEL trial. Cardiovasc Diabetol 2019; 18(1): 138.https://cardiab.biomedcentral.com/articles/10.1186/s12933-019-0942-x
[http://dx.doi.org/10.1186/s12933-019-0942-x] [PMID: 31640705]

[93] Endo A, Kuroda M. Citrinin, an inhibitor of cholesterol synthesis. J Antibiot (Tokyo) 1976; 29(8): 841-3.
[PMID: 791911]

[94] Brown MS, Goldstein JL. Familial hypercholesterolemia: defective binding of lipoproteins to cultured fibroblasts associated with impaired regulation of 3-hydroxy-3-methylglutaryl coenzyme A reductase activity. Proc Natl Acad Sci USA 1974; 71(3): 788-92.
[http://dx.doi.org/10.1073/pnas.71.3.788] [PMID: 4362634]

[95] Nair P. Brown and Goldstein: the cholesterol chronicles. Proc Natl Acad Sci USA 2013; 110(37): 14829-32.
[http://dx.doi.org/10.1073/pnas.1315180110] [PMID: 23980185]

[96] Alberts AW, Chen J, Kuron G, Hunt V, Huff J, Hoffman C, *et al.* Mevinolin: A highly potent competitive inhibitor of hydroxymethylglutaryl-coenzyme A reductase and a cholesterol-lowering agent. Proc Natl Acad Sci USA 1980; 77(7 II): 3957–61.
[http://dx.doi.org/10.1073/pnas.77.7.3957]

[97] Randomised trial of cholesterol lowering in 4444 patients with coronary heart disease: the Scandinavian Simvastatin Survival Study (4S). Lancet 1994; 344(8934): 1383-9.
[PMID: 7968073]

[98] Cannon CP, Braunwald E, McCabe CH, *et al.* Intensive *versus* moderate lipid lowering with statins after acute coronary syndromes. N Engl J Med 2004; 350(15): 1495-504.
[http://dx.doi.org/10.1056/NEJMoa040583] [PMID: 15007110]

[99] La Rosa JC, Conti CR. Intensive lipid lowering with atorvastatin in patients with stable coronary disease. ACC Cardiosource Rev J 2006; 15(5): 97-9.

[100] Amarenco P, Bogousslavsky J, Callahan A III, *et al.* High-dose atorvastatin after stroke or transient ischemic attack. N Engl J Med 2006; 355(6): 549-59.
[http://dx.doi.org/10.1056/NEJMoa061894] [PMID: 16899775]

[101] Downs JR, Clearfield M, Weis S, *et al.* Primary prevention of acute coronary events with lovastatin in men and women with average cholesterol levels: results of AFCAPS/TexCAPS. Air Force/Texas Coronary Atherosclerosis Prevention Study. JAMA 1998; 279(20): 1615-22.
[http://dx.doi.org/10.1001/jama.279.20.1615] [PMID: 9613910]

[102] Colhoun HM, Betteridge DJ, Durrington PN, *et al.* Primary prevention of cardiovascular disease with atorvastatin in type 2 diabetes in the Collaborative Atorvastatin Diabetes Study (CARDS): multicentre randomised placebo-controlled trial. Lancet 2004; 364(9435): 685-96.
[http://dx.doi.org/10.1016/S0140-6736(04)16895-5] [PMID: 15325833]

[103] MRC/BHF Heart Protection Study of cholesterol lowering with simvastatin in 20,536 high-risk individuals: a randomised placebo-controlled trial. Lancet 2002; 360(9326): 7-22.https://linkinghub.elsevier.com/retrieve/pii/S0140673602093273
[http://dx.doi.org/10.1016/S0140-6736(02)09327-3] [PMID: 12114036]

[104] Sever PS, Dahlöf B, Poulter NR, *et al.* Prevention of coronary and stroke events with atorvastatin in hypertensive patients who have average or lower-than-average cholesterol concentrations, in the Anglo-Scandinavian Cardiac Outcomes Trial--Lipid Lowering Arm (ASCOT-LLA): a multicentre randomised controlled trial. Drugs 2004; 64 (Suppl. 2): 43-60.
[http://dx.doi.org/10.2165/00003495-200464002-00005] [PMID: 15765890]

[105] Sever PS, Poulter NR, Dahlöf B, *et al.* Reduction in cardiovascular events with atorvastatin in 2,532 patients with type 2 diabetes: Anglo-Scandinavian Cardiac Outcomes Trial--lipid-lowering arm (ASCOT-LLA). Diabetes Care 2005; 28(5): 1151-7.
[http://dx.doi.org/10.2337/diacare.28.5.1151] [PMID: 15855581]

[106] Ridker PM, Danielson E, Fonseca FAH, *et al.* Rosuvastatin to prevent vascular events in men and women with elevated C-reactive protein. N Engl J Med 2008; 359(21): 2195-207.

[http://dx.doi.org/10.1056/NEJMoa0807646] [PMID: 18997196]

[107] Wanner C, Krane V, März W, *et al.* Atorvastatin in patients with type 2 diabetes mellitus undergoing hemodialysis. N Engl J Med 2005; 353(3): 238-48.
[http://dx.doi.org/10.1056/NEJMoa043545] [PMID: 16034009]

[108] Fellström BC, Jardine AG, Schmieder RE, *et al.* Rosuvastatin and cardiovascular events in patients undergoing hemodialysis. N Engl J Med 2009; 360(14): 1395-407.
[http://dx.doi.org/10.1056/NEJMoa0810177] [PMID: 19332456]

[109] Kjekshus J, Apetrei E, Barrios V, *et al.* Rosuvastatin in older patients with systolic heart failure. N Engl J Med 2007; 357(22): 2248-61.
[http://dx.doi.org/10.1056/NEJMoa0706201] [PMID: 17984166]

[110] Tavazzi L, Maggioni AP, Marchioli R, *et al.* Effect of rosuvastatin in patients with chronic heart failure (the GISSI-HF trial): a randomised, double-blind, placebo-controlled trial. Lancet 2008; 372(9645): 1231-9.
[http://dx.doi.org/10.1016/S0140-6736(08)61240-4] [PMID: 18757089]

[111] Baigent C, Blackwell L, Emberson J, *et al.* Efficacy and safety of more intensive lowering of LDL cholesterol: a meta-analysis of data from 170,000 participants in 26 randomised trials. Lancet 2010; 376(9753): 1670-81.
[http://dx.doi.org/10.1016/S0140-6736(10)61350-5] [PMID: 21067804]

[112] Mihaylova B, Emberson J, Blackwell L, *et al.* The effects of lowering LDL cholesterol with statin therapy in people at low risk of vascular disease: meta-analysis of individual data from 27 randomised trials. Lancet 2012; 380(9841): 581-90.
[http://dx.doi.org/10.1016/S0140-6736(12)60367-5] [PMID: 22607822]

[113] Baigent C, Landray MJ, Reith C, *et al.* The effects of lowering LDL cholesterol with simvastatin plus ezetimibe in patients with chronic kidney disease (Study of Heart and Renal Protection): a randomised placebo-controlled trial. Lancet 2011; 377(9784): 2181-92.
[http://dx.doi.org/10.1016/S0140-6736(11)60739-3] [PMID: 21663949]

[114] Herrington WG, Emberson J, Mihaylova B, *et al.* Impact of renal function on the effects of LDL cholesterol lowering with statin-based regimens: a meta-analysis of individual participant data from 28 randomised trials. Lancet Diabetes Endocrinol 2016; 4(10): 829-39.
[http://dx.doi.org/10.1016/S2213-8587(16)30156-5] [PMID: 27477773]

[115] Cannon CP, Blazing MA, Giugliano RP, *et al.* Ezetimibe added to statin therapy after acute coronary syndromes. N Engl J Med 2015; 372(25): 2387-97.https://www.scopus.com/inward/record.uri?eid=2-s2.0-84931411520&doi=10.1056%2FNEJMoa1410489&partnerID=40&md5=e05871501a2f038a737 89cac14db7ec2
[http://dx.doi.org/10.1056/NEJMoa1410489] [PMID: 26039521]

[116] Han BH, Sutin D, Williamson JD, *et al.* Effect of statin treatment vs usual care on primary cardiovascular prevention among older adults: The ALLHAT-LLT randomized clinical trial. JAMA Intern Med 2017; 177(7): 955-65.
[http://dx.doi.org/10.1001/jamainternmed.2017.1442] [PMID: 28531241]

[117] Abifadel M, Varret M, Rabès JP, *et al.* Mutations in PCSK9 cause autosomal dominant hypercholesterolemia. Nat Genet 2003; 34(2): 154-6.
[http://dx.doi.org/10.1038/ng1161] [PMID: 12730697]

[118] Cohen JC, Boerwinkle E, Mosley TH Jr, Hobbs HH. Sequence variations in PCSK9, low LDL, and protection against coronary heart disease. N Engl J Med 2006; 354(12): 1264-72.
[http://dx.doi.org/10.1056/NEJMoa054013] [PMID: 16554528]

[119] Grefhorst A, McNutt MC, Lagace TA, Horton JD. Plasma PCSK9 preferentially reduces liver LDL receptors in mice. J Lipid Res 2008; 49(6): 1303-11.
[http://dx.doi.org/10.1194/jlr.M800027-JLR200] [PMID: 18354138]

[120] Fonarow GC, van Hout B, Villa G, Arellano J, Lindgren P. Updated Cost-effectiveness Analysis of Evolocumab in Patients With Very High-risk Atherosclerotic Cardiovascular Disease. JAMA Cardiol 2019; 4(7): 691-5.
[http://dx.doi.org/10.1001/jamacardio.2019.1647] [PMID: 31166576]

[121] Leaf A. Historical overview of n-3 fatty acids and coronary heart disease. Am J Clin Nutr 2008; 87(6): 1978S-80S.
[http://dx.doi.org/10.1093/ajcn/87.6.1978S] [PMID: 18541598]

[122] Singh RB. GISSI-Prevenzione trial. Lancet 1999; 354(9189): 1556-7.
[http://dx.doi.org/10.1016/S0140-6736(05)76585-5] [PMID: 10551522]

[123] Lakshmanan S, Shekar C, Kinninger A, *et al.* Comparison of mineral oil and non-mineral oil placebo on coronary plaque progression by coronary computed tomography angiography. Cardiovasc Res 2020; 116(3): 479-82.
[http://dx.doi.org/10.1093/cvr/cvz329] [PMID: 31825484]

[124] Olshansky B, Chung MK, Budoff MJ, *et al.* Mineral oil: safety and use as placebo in REDUCE-IT and other clinical studies. Eur Heart J Suppl 2020; 22 (Suppl. J): J34-48.
[http://dx.doi.org/10.1093/eurheartj/suaa117] [PMID: 33061866]

CHAPTER 6

Diabesity and the Kidney

Mohamed E. Elrggal[1], Ahmed Elkeraie[1,2], Sol Carriazo[3], Hany Sawaf[4], Si Yuan Khor[5], Yasmine Elkeraie[1], Issa Haddad[5], Khaled Moustafa[2] and Mohamed Hassanein[6,*]

[1] *Kidney and Urology Center, Alexandria, Egypt*

[2] *Faculty of Medicine, Alexandria University, Alexandria, Egypt*

[3] *Department of Nephrology and Hypertension, IIS-Foundation Jimenaz-Diaz-UAM, Madrid, Spain*

[4] *Cleveland Clinic Foundation, Cleveland, OH, United States*

[5] *Michigan State University, East Lansing, MI, United States*

[6] *University of Mississippi Medical Center, Jackson, MS, United States*

Abstract: Diabetes Mellitus and obesity, now coined as "Diabesity", is a worldwide epidemic that imposes a huge burden on healthcare and society. Diabesity has been associated with poor outcomes and increased morbidity and mortality. The kidneys are a vulnerable target of diabesity. In this chapter, we discuss the epidemiology, pathophysiology, and treatment of diabesity–induced kidney disease. We specifically focus on the therapeutic targets and pharmacological management of diabesity-related kidney diseases.

Keywords: Chronic kidney disease, Diabesity, Diabetes, Diabetic kidney disease, Kidney failure, Obesity.

INTRODUCTION

Diabetes Mellitus (DM) and obesity, now coined as "Diabesity", is a worldwide epidemic that imposes a huge burden on healthcare and society [1]. Five million deaths in 2015 were attributed to DM in people aged 20–79 years, representing 12.8% of the global all-cause mortality. In 2010, the prevalence of diabetes worldwide was 284 million people, which represented nearly 6.4% of the whole world population and is estimated to reach 642 million by 2040 [2, 3].

* **Corresponding author Mohamed Hassanein:** University of Mississippi Medical Center 2500 North State Street, Jackson 39216, MS, USA; Tel: +16019845670; E-mail: mhassanein@umc.edu

Shazia Anjum (Ed.)

Insulin resistance is the cornerstone of the pathophysiology of diabesity. For every kilogram rise in body weight, there is an increased risk of diabetes by 4.5%. Poor dietary habits, lack of exercise, and other risk factors lead to hyperinsulinemia, insulin resistance (IR), and atherogenic dyslipidemia, *i.e.*, hypertriglyceridemia, low high-density lipoprotein (HDL-C), and increased low-density lipoprotein (LDL-C).

Diabesity predisposes to cardiovascular morbidity and other comorbidities, such as hypertension, endothelial dysfunction, metabolic syndrome, and obstructive sleep apnea. Moreover, diabesity is linked with polycystic ovarian syndrome and various malignancies, such as breast, endometrial, and prostate cancer [3, 4].

The kidney is the most important target of microvascular damage in DM. Diabetic kidney disease (DKD) is the leading cause of end-stage kidney disease (ESKD) worldwide. It is now considered as a "medical catastrophe of worldwide dimension" [5]. Diabetes that affects the kidneys used to be known as Diabetic Nephropathy, however, DKD is now the new term used to encompass a whole spectrum of nephro-pathology induced by DM, since it has been introduced by the Kidney Disease Outcomes Quality Initiative (KDOQI) in 2007 [6].

Genetically speaking, DM can be categorized into a monogenic form, including neonatal and maturity-onset diabetes of the young (MODY), and a polygenic form that includes classic Type 1 Diabetes Mellitus (T1DM) and Type 2 Diabetes Mellitus (T2DM). Studies have shown that DKD in T2DM has not been strictly linked to poor glycemic control as it was more likely to develop in patients with a strong family history of cardiovascular disease [7].

Obesity presents a systemic pro-inflammatory state that promotes IR and DM [8]. Obesity-associated nephropathy is characterized by increased kidney weight and hypertrophy of individual nephrons, increased glomerular size, and reduced glomerular density in the cortex, as well as the number of glomerular capillaries [9].

It's always challenging in patients suffering from diabetes and kidney disease to distinguish between those wit non-diabetic CKD [10]. Therefore, despite numerous studies relying on nephro-pathology to differentiate DKD from non-DKD, it remains sometimes difficult to determine the exact incidence of DKD.

In addition to the current advances in understanding DKD, there are some concerns that physicians have regarding optimizing the diagnostic process and future propositions in its management. The natural course of DKD could be divided into five stages: increased glomerular filtration rate (GFR) initially with hyperfiltration, the 'silent' phase, the 'incipient' phase, the 'overt' phase, and eventually the development of ESKD. Nevertheless, not all patients obviously follow the same course of complications [11].

DKD screening should be done yearly in T1DM starting 5 years after diagnosis and at the time of diagnosis for all patients with T2DM then annually thereafter. Diabetic retinopathy is strongly suggestive of DKD in the presence of albuminuria. To confirm the diagnosis of DKD, albuminuria or reduced estimated GFR (eGFR) should be present in two abnormal measurements at least 3 months apart [12].

The atypical presentation that may denote non-diabetic kidney disease includes sudden onset of low eGFR, rapidly decreasing eGFR, an abrupt increase in albuminuria, development of nephrotic-range proteinuria, development of nephritic syndrome, refractory hypertension, signs or symptoms of another systemic disease, and > 30% eGFR decline within 2–3 months of initiation of a renin-angiotensin system inhibitor [12, 13].

According to the American Diabetes Association, glycemic targets should be tailored to age and other comorbidities. Strict glycemic control, such as hemoglobin A1c (HbA1c) <6.5%, is important for young patients with early diabetes, and those who have not yet developed complications. On the other hand, HbA1c targets up to 8% are allowed for patients with longstanding DM, older age, micro-and macrovascular complications, and limited life expectancy. Similarly, the National Kidney Foundation (NKF)–KDOQI and the Kidney Disease Improving Global Outcomes (KDIGO) guidelines recommend a target HbA1c of about 7% to prevent or delay the progression of the microvascular complications of diabetes. However, patients at risk for hypoglycemia, should not target less than that.

Conventional therapy of DKD includes good hyperglycemic control, control of hypertension and hyperlipidemia, antiproteinuric drugs, and close monitoring for micro and macrovascular complications with appropriate management to slow their progression.

Anti-diabetic medications also have an impact on weight in addition to glycemic control. Recent clinical trials on patients with DKD revealed that Sodium-Glucose co-transporter-2 inhibitors (SGLT2i) and glucagon-like peptide-1 receptor agonists (GLP-1 RA) were capable of improving long-term kidney-related outcomes of DKD in the outpatient clinic [5, 14]. Newer agents to treat obesity are emerging, with efficacy and safety being tested in randomized controlled trials.

Epidemiology and Genetic Aspects Shared in Diabesity and Kidney Diseases

The connections between DM, CKD, and obesity can be less clear than one might assume. On a global scale, the prevalence of DM has been noted to be closely related to the prevalence of obesity. In places where obesity affects <10% of the population; DM will be found in < 6% of that population. On the other hand, obesity affects >30% of the population which has a DM rate of > 14% [15]. According to the US renal data system (USRDS), those surveyed in 2015 – 2018 had CKD at a rate of 14.9%, which was stable compared to the previous report. It was also noted that 13.1% of the US population had a diagnosis of DM compared to about a third of patients with CKD who were also diagnosed with DM [16].

Unfortunately, even when we have several patients who suffer from both DM and CKD, it has proven challenging to determine if these patients with DM have CKD secondary to diabetic kidney disease *versus* being diabetics with non-diabetic CKD [10]. There have been biopsy studies that were used to differentiate DKD from non-DKD. These rates have been found to vastly differ when all countries are considered; when evaluating, just two studies in the US showed that the rate of pure DKD amongst diabetics was 28% [17] to 37% [18]. Given the nature of the studies, it is also possible that these rates overestimate patients with DKD [10]. So even though we know the number of patients with DM and the number of patients with CKD, it has proven challenging to determine the incidence of DKD.

As for the questions related to the inheritance and the genetics of DKD; experts have tried to find answers for decades. It was noted and even published in the New England Journal of Medicine back in the late 1980s that DKD occurs in familial clusters [19]. This familial clustering of DKD was noted in families throughout different populations [20], including in the US [21, 22], Italy [23], Brazil [24] and India [25].

When comparing different ethnicities to one another, it has been found in studies from different countries that a significant difference exists in the rates of DKD. In the United Kingdom (UK), a 5-year retrospective cohort study of 3855 people with DM revealed that individuals of South Asian and African descent had a more rapid decline in kidney function when compared to whites [26]. There have been similar findings in the US, where African Americans have a 2-4-fold greater incidence of ESKD in comparison to the white population [27], and when looking specifically at DM-related ESKD, it was found that African Americans, Hispanics, and Native Americans have a higher incidence of ESKD related to DM [28]. There have also been recent publications showing that women with DM are more likely to develop diabetic kidney disease than men [29].

Even though there have been clear findings that suggest a genetic predisposition of DKD [30], things quickly get complicated when trying to find specific loci and define them. Some of the landmark studies looking into the genetics of these topics include the Genetics of Nephropathy: An International Effort (GENIE) consortium, the FIND study, and the Genetics of Kidneys in Diabetes (GoKinD) study. The GoKinD study was the earliest of the three and included 3075 participants to identify genes involved in DKD [31]. This was followed by the FIND study, which consisted of a genome-wide scan of more than 5,500 single nucleotide polymorphisms done on 1,235 nuclear and extended pedigrees of 3,972 participants with DM [32]. The GENIE consortium was a meta-analysis of genome-wide association studies (GWAS) of T1DM composing about 2.4 million single nucleotide polymorphisms in 6,691 individuals [33]. These different studies focus on the genetics of DKD and found evidence of linkage, but it remains unclear as to what can be done with these possible linkages that were detected.

To further complicate the topic of the genetics behind DKD; it was even found that there is evidence of epigenetic factors that can play a role in DKD. Changes in methylation can cause kidney damage to continue to worsen even after a patient achieves tighter glycemic control [34]. Furthermore, there have been studies where the methylation changes associated with different DKD phenotypes were defined, showing different patterns of methylation that can cause these different phenotypes [35].

Looking into a potential genetic link between DM and obesity, there have not been robust or convincing findings on this topic. It has proven difficult to know if the link between T2DM is driven by genetic causes or common risk variants shared between these populations [36]. There have been 185 identified loci that are associated with obesity traits identified in large-scale GWAS [37]; however, studies looking into these links have generally just looked into such high-risk alleles in small samples which cannot provide insight into the loci that they are identifying [38]. FTO gene is the strongest known obesity-risk locus described in the literature, which is known to have a well-defined interaction with physical activity and obesity [39]. However, determining if variants in this gene have a relationship with T2DM has not been consistently described through different populations [40].

Even though huge strides have been made in further understanding the genetics and the epigenetics behind DKD, there remains a lot to be done. DM is projected to affect 693 million adults by 2045 [41], and further research is required if we hope to reach our goal of identifying patients at risk for DKD and how that risk can be mitigated.

Pathophysiologic Mechanisms Cross-linking Diabesity and Kidney Diseases

The strong association between obesity and DM is well described; however, pathophysiological mechanisms underlying the role of diabesity in the development of CKD are still poorly understood. To comprehend the pathogenesis of diabesity, we will review the mechanisms of each condition and their impact on kidney diseases.

Obesity

Obesity presents a systemic pro-inflammatory state that promotes insulin resistance (IR) and DM [8]. A variety of population-based studies have reported associations between obesity and the development and progression of CKD [42 - 45]. Poor kidney-related outcomes in patients with obesity persist after adjustments for possible confounders such as high blood pressure and DM, pointing out possible different mechanisms [46].

Obesity-associated nephropathy is characterized by increased kidney weight and hypertrophy of individual nephrons, increased glomerular size, and reduced glomerular density in the cortex, as well as the number of glomerular capillaries [9]. The main clinical sign is albuminuria, which has been related to focal segmental glomerulosclerosis (FSGS), with diverse intensity according to the obesity stage. FSGS may also be accompanied by progressive interstitial fibrosis as loss of kidney function goes on.

Structural changes in the obese kidney leading to a higher risk of CKD include [1] ectopic lipid accumulation and higher deposition of kidney sinus fat and [2] an increase in the intraglomerular pressure leading to higher glomerular permeability due to the glomerular filtration barrier injury, resulting in glomerulomegaly and focal or segmental glomerulosclerosis [47].

Other effects of obesity on kidneys include malignancies and nephrolithiasis [46], which is related to a lower pH and increased urinary oxalate, uric acid, sodium, and phosphate excretion, among others [48]. Furthermore, increased adiposity has been related to kidney damage, induced by the endocrine activity of the adipose tissue *via* the production of different inflammatory mediators, including adiponectin [49], leptin [50], resistin [51], and visfatin [52]. This leads to inflammation, oxidative stress, activation of the renin-angiotensin-aldosterone system (RAAS), abnormal lipid metabolism, and increased production of insulin with further IR [46].

Further mechanisms involve the downregulation of corticosteroid 11 B-dehydrogenase isozyme 2 in the kidney, causing a rise in cortisol levels, and worsening hypertension through activation of the mineralocorticoid receptors [53, 54]. Additionally, obesity is associated with increased expression of neprilysin which might contribute to impaired urinary sodium excretion in obese patients [55].

Mechanical mechanisms have also been described. Compression of the kidney parenchyma by excess adipose tissue deposition might promote sodium reclamation by slowing the peritubular capillary flow and enhancing tubular solute reabsorption by the countercurrent multiplier [56]. In the same way, compression of the renal vein might cause retrograde pressure transmission to glomerular capillaries leading to glomerular hyperfiltration and glomerulomegaly. This situation could contribute to the development of glomerulosclerosis and tubulointerstitial fibrosis.

However, despite current knowledge, the precise mechanisms involving obesity-induced CKD remain unclear. Most obese individuals never develop CKD, suggesting that increased weight solely might not be sufficient to cause kidney damage.

Diabetes

The pathogenesis of DKD has been associated with functional and structural changes in the kidney in response to hyperglycemia [57]. These structural changes occur in multiple kidney compartments [12]. Molecular and epigenetic pathways have been suggested to be responsible for the early kidney functional loss in DM.

It is known that hyperglycemia might induce dysfunction of kidney and vascular cells. Therefore, chronic hyperglycemia induces activation and changes of metabolic pathways and hemodynamic dysfunction. One of the earliest lesions observed in DKD is the thickening of the glomerular basement membrane (GBM) visualized on electron microscopy. T1DM mainly affects the glomeruli, with thickening of the GBM and mesangial expansion, especially at the early stages. The kidney tubules, podocytes, arterioles, and interstitium are also affected, but at later stages [58 - 60]. Whereas, in patients with T2DM, two distinctive patterns of glomerular pathology are observed; nodular and non-nodular [61]. Nodular type glomerulosclerosis, also known as Kimmelstiel-Wilson nodules, was considered the only specific feature of DKD in the past [62]. These lesions consist of areas of marked mesangial expansions forming large round fibrillar mesangial zones with palisading of mesangial nuclei around the periphery of the nodule and compression of the associated glomerular capillaries [63]. These nodules are frequently described in advanced DKD, associated with heavy proteinuria and decreased GFR [60]. Currently, DKD classification includes 4 classes of glomerular lesions.

Metabolic changes that modify homeostasis, and hemodynamics and promote inflammation and fibrosis in early DM include hyperglycemia, hyperaminoacidemia, and hyperperfusion [12, 64]. Furthermore, the increased proximal tubular reabsorption of glucose through Sodium-Glucose cotransporter 2 (SGLT2) receptor, with a subsequent decrease in distal delivery of solutes, especially sodium chloride, to the macula densa is central in the pathogenesis of the disease [12, 64, 65]. High intraglomerular pressure and glomerular hyperfiltration are produced secondary to the decrease in tubuloglomerular feedback, dilating the afferent arteriole to increase glomerular perfusion and vasoconstriction of efferent arteriole secondary to local angiotensin II (Ang-II) production [64, 66].

The RAAS, protein kinase C (PKC), mitogen-activated protein kinase (MAPKs), reactive oxygen species (ROS), advanced glycation end products (AGE) formation, connective tissue growth factor (CTGF), activation of transforming growth factor- β 1 (TGF- β1), are main pathways participating in the progression of DKD. These different pathways and mediators overlap each other, making it even more complex to understand the nature of the disease [57].

High glucose stimulates renin and Angiotensin II (Ang-II) synthesis in mesangial cells, which might contribute to kidney injury by increasing glomerular capillary pressure and permeability, synthesis of cytokines and extracellular matrix, along with the promotion of macrophage infiltration, inflammation, renal cell proliferation, and hypertrophy [67, 68]. Chronic hyperglycemia could lead to oxidative stress, and increase Ag-II levels, PKC activation, and TGF-B expression, which stimulates the synthesis of the mesangial matrix and induce fibrotic processes in the tubular interstitium. Increased Ang-II and TGF-β elevate kidney ROS production through activation of NADPH oxidase. It is noteworthy to mention that oxidative stress is a common product of many pathways involved in the pathogenesis of DKD [69 - 71].

Some inflammatory factors such as IL-6, tumor necrosis factor (TNF-alpha), TGF-β 1, and IL-18, involved in the development and progression of DKD, are reported to be increased in the circulation [72]. IL-6 levels correlate with the severity of albuminuria and structural changes.

Secretion of proinflammatory and fibrogenic cytokines can cause direct damage and then facilitate the epithelial-to-mesenchymal transition (EMT) process [73]. Other elements, including transcription factors such as nuclear factor-kβ (NF-kβ TGF-β), Nrf2, as well as the activation of PKC-alpha and PKC- TGF-β isoforms, which are associated with increased NADPH activity, also participate in the inflammatory process that contributes to kidney injury. PKC-β activation could lead to kidney impairment through increased expression of p47 phox, Nox.2, Nox-4, endothelin-1, VEGF, TGF-β, CTGF, and oxidant production [69, 74].

In summary, the interaction between oxidative stress and inflammation plays a central role in the pathogenesis and progression of DKD. Oxidative stress induces kidney injury by intra and extracellular oxygen-derived radicals and inflammation [75]. While inflammation could also be produced by NF-k β activation due to an increase in AGE leading to ROS production. At the same time, ROS production activates NF-kβ and inflammatory cells.

Diabesity

Obesity frequently precedes T2DM, leading to metabolic syndrome and hypertension. Although molecular mechanisms contributing to the development of CKD and obesity are not the same, both conditions are accompanied by glomerular hypertrophy and transient hyperfiltration, mesangial matrix expansion, and glomerular basement membrane thickening resulting in albuminuria [43, 76, 77]. Along with kidney inflammation [78]and oxidative stress [79], these changes may lead to fibrosis and eventual decline in GFR [80].

In the context of obesity, IR results from increased free fatty acids (FFAs) levels, pro-inflammatory cytokines, and diacylglycerol (DAG), which inhibit phosphorylation of the insulin receptor substrate 1 (IRS-1) in phosphorylation domains (serine/threonine), preventing the propagation of signals to the translocation of the glucose transporter-4 (GLUT4) to the plasma membrane [57]. In this situation, the interaction between insulin and its receptor is affected, leading to decreased insulin-dependent glucose uptake, and eventually to hyperglycemia and hyperinsulinemia.

Overall, in obese diabetic patients, the increased accumulation of adipose tissue produces adipocytes stress through hyperplasia and hypertrophy, leading to subclinical inflammation, hypoxia, increased macrophage infiltration, and release of proinflammatory cytokines as tumor necrosis factor-alpha (TNF-alpha), Interleukin-6 (IL-6), and Interleukin-1 (IL-1) aggravating IR. TNF-alpha stimulates the secretion of other cytokines and chemokines and activates NF-kβ, facilitating chronic hyperglycemia, by affecting glucose uptake and promoting IR.

Glucose uptake, then, occurs in cells non-dependent on GLUT4 (non-insulin-dependent), such as kidney cells, that express GLUT1 and GLUT2 as glucose transporters, which do not regulate glucose entry into cells, generating glucotoxicity [57]. In this situation, these transporters become overexpressed, increasing the entry of glucose into kidney cells, as occurs with the high-affinity glucose transporter GLUT2, stimulated by hyperglycemia, SGLT 1, and SGLT2, responsible for tubular reabsorption [81, 82]. SGLT2 and SGLT1 are in charge of glucose transport on the apical membrane of kidney proximal tubule cells (KPTCs), requiring the presence of a sodium gradient provided by Na+/K+ ATPase at the basolateral membrane. Na+/K+ ATPase activity accounts for 40% of the energy expenditure in all kidneys [83]. In DM, SGLT2 protein levels in KPTCs are reported to be increased [84], however, others documented an opposite trend in its gene expression in different sets of T2DM patients without kidney disease [85, 86].

SGLT2 reduces sodium delivery to the macula densa, reducing vasoconstrictor glomerular feedback to the afferent arteriole [87]. To prevent the accumulation of toxic glucose metabolites in KPTCs, SGLT2 exerts all reabsorbed glucose. KPTCs are tuned to gluconeogenesis instead of glycolysis and metabolize little glucose and rely on nutrients such as fatty acids, lactate, ketone bodies, and glutamate [83, 88]. In this process, mTORC1, an important regulator of amino acid transport and glycolysis, may enhance kidney fibrosis and injury and avoid the kidney-protective effects of SGLT2i in DM [83].

As seen, in all this process, the KPTCs play an important role. KPTCs are exposed to extensive amounts of nutrients, through reabsorption from the lumen and *via* systemic circulation. Their strategic localization, near the glomerulus, where the urinary filtrate is formed, gives them susceptibility to nutrient overload with high glucose and/or high-fat diet, as occurs in DM and obesity [83]. These cells retain insulin sensitivity during hyperinsulinemia, contributing to increased sodium reclamation in the kidney in obesity. On the other hand, podocytes become insulin resistant, facilitating dedifferentiation and the loss of selectivity across the GBM [89].

During obesity, the accumulation of visceral adipose tissue leads to adipocyte stress, pro-inflammatory signaling, and lipolysis [90]. Ectopic lipid accumulation in the kidney and toxic levels of intracellular lipid metabolites can drive oxidative stress, induce IR in podocytes, and lead to associated glomerular barrier dysfunction [87]. This lipid accumulation is also frequently seen in the liver of nearly 50-75% of patients with diabesity. Non-alcoholic fatty liver disease (NAFLD) is now recognized as a risk factor for CKD. There is a close relationship between IR and NAFLD, with the prevalence of NAFLD being 5-fold higher in patients with DM compared to those without [91].

IR increases the flux of FFAs from adipocytes to the liver, promoting hepatic lipid accumulation and liver injury. The development of diabesity contributes to the development of steatosis, hepatic inflammation, and fibrosis [76, 92]. There is evidence associating NAFLD as an independent risk factor to the development and progression of CKD [93], with a positive correlation with severity [94]. Several factors, such as DM and obesity, could also contribute to mechanisms linking NAFLD and CKD [95, 96].

Even though the effect of DM and obesity on kidney disease is well known, more data is needed to improve our understanding of diabesity in kidney damage, and possible therapeutic approaches.

Non-Pharmacological Management

As obesity is a disease mainly due to a consequence of an imbalance between overconsumption and underutilization of nutritional energy intake; lifestyle modification is the very first step in the management of diabesity to disrupt this pathological state of disequilibrium [97]. Intense lifestyle changes involve regular exercise, dietary therapy, and behavioral modifications to achieve weight reduction. Weight loss of at least >5% should be the targeted goal [98]. Look AHEAD Trial demonstrated the benefits of weight loss in patients with DM; it showed that weight loss of 5-10% reduced the HbA1c levels, improved cardiovascular disease risk factors, and decreased the use of anti-hyperglycemic, anti-hypertensive and lipid-lowering agents after 1 year [99]. Weight reduction can be achieved with exercise, dietary therapy, and behavioral modifications.

A. Exercise

Increased energy expenditure *via* physical activity is a strong predictor of weight loss maintenance. Although aerobic exercise is the best mode of exercise to decrease fat mass, multicomponent exercise programs that consist of aerobic and resistance training are preferred [99]. Moreover, exercise increases lean mass, and improves glucose tolerance and insulin sensitivity. To prevent weight gain and improve cardiovascular health, it is advised to exercise at least 30 minutes or more, 5-7 days a week [100]. In the context of exercise in achieving weight loss, it appears to be a dose effect, and thus to achieve significant weight reduction without energy-restricted diets, a higher amount of physical activity is required [100].

B. Dietary Therapy

Studies have also shown that calorie-restricted diets have synergistic effects on weight loss when it is combined with exercise; these include balanced low-calorie, low-fat/low-calorie, moderate-fat/low-calorie, the low-carbohydrate or Mediterranean diet [101, 102]. Dietary adherence is the key regardless of the type of diet regimen chosen [103]. Hence, the diet should be tailored to reduce energy intake below energy expenditure according to an individual's preference, lifestyle, and underlying medical comorbidities [101, 102]. Dietary counseling may also facilitate weight reduction, particularly during the first year of treatment.

C. Behavioral Modifications

These intense lifestyle modifications have limitations and are difficult to sustain, particularly in real-life settings; thus, implementation of such interventions can be extremely challenging. Behavioral therapy is fundamental in the treatment of diabesity. The US Preventive Services Task Force (USPSTF) recommends that all adults with BMI ≥30 be offered intensive, multicomponent behavioral therapy [104]. The goal of behavioral therapy is to facilitate patients in making sustainable long terms lifestyle modifications in terms of modifying and monitoring food intake, modifying physical activities, and controlling cues and stimuli in the environment that trigger eating and overeating [105].

Pharmacological Treatment for the Management of Diabesity and Related Kidney Disease

Weight-losing agents generally are either true anti-diabetic agents that cause weight loss, or drugs that cause weight loss and hence help in managing DM and insulin resistance states.

Anti-diabetic medications have different consequences on weight. Some medications are considered weight neutral, like dipeptidyl peptidase-4 (DPP-4) inhibitors and alpha-glucosidase inhibitors. Metformin is considered weight neutral, though it may be associated with slight weight loss, while GLP1-RA and SGLT2 inhibitors are associated with weight loss. Weight gain is an adverse effect of many medications, including insulin, sulfonylureas, and thiazolidinediones [106]. Preference should be given to anti-diabetic therapies that do not increase the weight of the patient and ideally support weight reduction.

Several anti-obesity drugs have been marketed worldwide during the last century, but most of them were withdrawn later because of serious adverse effects (AEs), such as psychiatric disturbances (*e.g.*, clobenzorex, fenbutrazate, fenproporex, mazindol, and rimonabant), cardiovascular toxicity (*e.g.*, amfepramone, aminorex, benfluorex, chlorphentermine, dexfenfluramine, norpseudoephedrine, and sibutramine), and abuse potential (*e.g.*, amphetamine, methamphetamine, phendimetrazine, phentermine, pipradrol, and Pyrovalerone) [107]. Lorcaserin was withdrawn by FDA in 2020 due to the potential risk of cancer [108].

Current FDA-approved medications for obesity are orlistat, phentermine, topiramate fixed combination, naltrexone, bupropion fixed combination, liraglutide, and semaglutide. This review will cover anti-diabetic medications with potential weight loss effects, anti-obesity medications, as well as some novel agents (Table **1**).

Table 1. The pharmacological and non-pharmacological approach in the management of diabesity and kidney disease.

Approach	Examples
Non- pharmacological approaches	
	1. Exercise 2. Dietary therapy 3. Behavioral modification
Pharmacological approaches	
a. Anti-diabetes medications	1. Glucagon-like peptide-1 receptor agonists (GLP1-RA) 2. Sodium-Glucose co-transporter-2 inhibitors (SGLT2i) 3. Metformin
b. Anti-obesity medications	1. Orlistat 2. Phentermine/topiramate 3. Naltrexone/bupropion
Surgical options (bariatric surgery)	
	1. Sleeve gastrectomy 2. Roux-en-Y gastric bypass 3. Gastric banding
Future pipeline treatments	
	1. Setmelanotide 2. Tirzepatide 3. Cotadutide 4. Amylin analogs (Cagrilintide, Davalinitide) 5. Leucine/Metformin/Sildenafil Combination 6. Beloranib 7. Cetilistat 8. anti-obesity vaccines (anti-Ghrelin and Somatostatin antibodies)

ANTI-DIABETIC MEDICATIONS

GLP-1 Receptor Agonists

The observation that oral glucose administration produced a greater insulin response compared to an intravenous glucose infusion, led to connecting the gastrointestinal system and insulin secretion back in the 1960s [109]. This 'incretin effect' is primarily mediated by two insulinotropic gut hormones, GLP-1 and gastric inhibitory polypeptide (GIP), and accounts for approximately 50% - 70% of the total insulin secreted following oral glucose administration [110]. The incretin effect is greatly impaired in patients with T2DM, and GLP-1 secretion is noticeably deficient.

GLP-1 receptors are found in the pancreas, stomach, intestine, heart, kidney, peripheral and central nervous system. Several responses occur after activation of GLP-1 receptors, including glucose-dependent stimulation of pancreatic insulin secretion and inhibition of extensive glucagon secretion. Also, activation of GLP-1 receptors regulates appetite and caloric intake, slows gastric emptying, and promotes weight loss.

As native GLP-1 is rapidly degraded by endogenous DPP-4 enzyme promoting a short half-life of 1.5 - 2 minutes, several stable analogs were produced by recombinant DNA technology.

Liraglutide

Liraglutide is a GLP-1 analog, with a prolonged plasma half-life of 13 hours, allowing for once-daily administration. Liraglutide does not inhibit cytochrome P450 enzymes and has no clinically significant drug interactions. Although elimination through the kidney does not appear significant, liraglutide's area under the curve was lower in patients with mild to severe kidney impairment. Data on hepatic impairment is limited. Caution should be utilized in patients with kidney or hepatic impairment. There have been post-marketing reports of acute kidney injury with GLP-1 receptor agonists, including liraglutide, however, some of these reports were in patients with underlying kidney disease, and the majority occurred in volume-depleted patients [111].

Liraglutide was associated with a significant excess of weight loss of 5.3 kg (95% CI, -6.06 to -4.52 kg) compared to placebo at 1 year, though it was shown to have high odds of being discontinued because of AEs (OR, 2.95; 95% CI, 2.11-4.23) [112]. The most frequent side effects were nausea (25.0%), vomiting (12.2%), diarrhea (11.6%), constipation (11.0%), and dyspepsia (6.4%). Elevations of liver enzymes, hepatitis, angioedema and anaphylactic reactions, rash, and pruritus were also reported. Liraglutide was associated with dose-dependent and treatment-duration-dependent thyroid C-cell tumors in rats and mice. It is contraindicated in patients with a personal or family history of medullary thyroid carcinoma or patients with multiple endocrine neoplasia syndrome type 2. Based on post marketing reports, acute pancreatitis, including fatal and non-fatal hemorrhagic or necrotizing pancreatitis, has been observed in patients treated with liraglutide. Patients should be observed carefully for signs and symptoms of pancreatitis (including persistent severe abdominal pain, sometimes radiating to the back and which may or may not be accompanied by vomiting). If pancreatitis is suspected, liraglutide should promptly be discontinued and appropriate management should be initiated. If pancreatitis is confirmed, liraglutide should not be restarted [111]. However, long-term trials suggest that the risk of pancreatitis is not significantly increased with GLP-1 receptor agonists, whereas a significantly increased risk of cholelithiasis was detected [113].

In the LEADER trial, 3.1% of liraglutide-treated patients *versus* 1.9% of placebo-treated patients reported an acute event of gallbladder diseases, such as cholelithiasis or cholecystitis. Most events required hospitalization or cholecystectomy [114]. If cholelithiasis is suspected, gallbladder studies and appropriate clinical follow-up are indicated.

Dulaglutide

The half-life of dulaglutide is approximately five days, which allows for once-weekly administration. It is approved for the treatment of adults with T2DM. The effect on body weight appears to be dose related. The AWARD-4 trial compared once-weekly dulaglutide *versus* bedtime insulin glargine. At 52 weeks, body weight decreased in patients in the dulaglutide 1·5 mg group and slightly increased in those in the 0.75 mg group. These changes differed significantly from the increase in weight noted with glargine, with clinically relevant between-group differences [115]. In the AWARD-5 trial, the mean change in body weight after 52 weeks was significantly better with dulaglutide compared with sitagliptin (dulaglutide 0.75 mg, -2.6 and dulaglutide 1.5 mg, -3 kg *versus* -1.53 kg with sitagliptin) [116]. In an exploratory analysis of the AWARD-11 trial, higher dulaglutide doses were associated with numerically greater weight reduction compared with 1.5 mg in each baseline BMI (1.5 mg, −3.1 kg; 3.0 mg, −4.0 kg [P = .001]; 4.5 mg, −4.7 kg [P < .001]) [117].

Semaglutide

Semaglutide is a long-acting GLP-1 analog that can be administered once weekly subcutaneous or once-daily oral dose. It received FDA approval in 2017 as an adjunct to diet and exercise to improve glycemic control in adults with T2DM. Recently in June 2021, the FDA approved subcutaneous injection use for chronic weight management in adults with obesity or overweight with at least one weight related condition (such as high blood pressure, T2DM, or high cholesterol), to be used in conjunction with a reduced-calorie diet and increased physical activity.

Semaglutide has been shown to promote weight loss in individuals with or without DM. In the STEP1 trial, 1,961 individuals with obesity received a high dose of semaglutide or placebo. At 68 weeks, semaglutide recipients had lost

12.4% of their initial body weight, after adjusting for the placebo response. Of semaglutide recipients, 69% lost more than 10% of their body weight. More participants in the semaglutide group discontinued treatment due to gastrointestinal AEs compared with those in the placebo group (4.5 *versus* 0.8 percent) [118]. Common side effects included nausea and diarrhea, typically transient and mild-to-moderate in severity.

Results from the STEP2, STEP3, and STEP4 trials showed placebo-adjusted weight loss of 6.2–11.5% [119 - 121]. The average Semaglutide placebo-adjusted weight loss across the four trials was just over 10%. A shorter duration of treatment is associated with weight regain. In the STEP 4 trial, participants continuing semaglutide after 20 weeks of initial therapy continued to lose weight, while those who switched to placebo regained weight over the subsequent 48 weeks [121]. Compared to dulaglutide, bodyweight reductions were greater with semaglutide across all baseline BMI subgroups in the SUSTAIN 7 trial [122]. Semaglutide 1.0 mg was superior to liraglutide 1.2 mg in reducing HbA1c and body weight in the SUSTAIN 10 trial. Safety profiles were generally similar, except for higher rates of gastrointestinal AEs with semaglutide [123]. Yet, there are no available head-to-head trials that compare the efficacy of once-daily liraglutide 3.0 mg *versus* once-weekly semaglutide 2.4 mg. Once-daily subcutaneous administration of semaglutide (0.05 mg, 0.1 mg, 0.2 mg, 0.3 mg, or 0.4 mg; initiated at 0.05 mg per day and incrementally escalated every 4 weeks) was compared with liraglutide (3.0 mg; initiated at 0.6 mg per day and escalated by 0.6 mg per week). Mean body weight reductions for 0.2 mg or more of semaglutide *versus* liraglutide were all significant (–13.8% to –11.2% *vs.* –7.8%) [124].

The initial dose is 0.25 mg once weekly for four weeks. The dose is increased at four-week intervals (0.5, 1, 1.7, 2.4 mg) to the recommended dose of 2.4 mg once weekly. If dose escalation is not tolerated due to side effects (*e.g.*, nausea, vomiting), the increase in dose can be delayed by another four weeks [125].

Dosage adjustment is not necessary for patients with kidney impairment. The SUSTAIN 6 trial showed that semaglutide reduced major cardiovascular disease events in adults with T2DM and established cardiovascular disease or chronic kidney disease. Data indicate that 0.5-1.0 mg weekly can safely be used at an eGFR of less than 30 $mL/min/1.73m^2$ for up to 2 years [126]. However, the dose of semaglutide used was lower than the dose recommended for weight loss (0.5 and 1.0 *versus* 2.4 mg).

Semaglutide is contraindicated during pregnancy and in patients with a personal history of pancreatitis or a personal or family history of medullary thyroid cancer or multiple endocrine neoplasia 2A or 2B. Rare cases of angioedema and anaphylaxis have been reported with semaglutide. Patients with diabetic retinopathy should be monitored for complications [125].

With the increase in the prescription of GLP1-RA to treat obesity, it should be noted that prescription has to be under medical supervision to prevent inappropriate use of these drugs due to a heightened desire to lose weight, which can lead to unfortunate outcomes.

SGLT2 Inhibitors

SGLT2 inhibitors, such as dapagliflozin, empagliflozin, and canagliflozin, block glucose reabsorption from the kidney tubules, resulting in glycosuria. They were primarily used as anti-diabetic medications in patients with T2DM. However, they were found to have pluripotent effects. SGLT2 inhibitors decrease weight. In a meta-analysis, weight loss was significant with SGLT2 inhibitors compared to placebo (for 1-year result, weighted mean differences: -2.477; 95% CI: -2.568 to -2.385; for 2-years result, weighted mean differences: -2.990; 95% CI: -3.642 to -2.337) [127]. Besides, they promote weight loss, even in individuals without DM [128]. Moreover, a recent meta-analysis emphasized that all SGLT2 inhibitors were effective at reducing weight, with canagliflozin being the most effective [129].

The amount of glucose loss in urine is approximately 70–90 g/day, equivalent to 300 kcal/day [130]. However, weight loss with SGLT2 inhibitor was found to be lower than expected [131]. This discrepancy between the anticipated weight loss and actual weight reduction may be explained by adaptive changes, such as compensatory hyperphagia, increased gluconeogenesis, and a shift toward fatty acid utilization as a metabolic substrate [130, 132]. Thus, combining an appetite suppressor, such as phentermine, with an SGLT-2 inhibitor can improve weight loss, by counteracting the possible metabolic compensatory mechanisms. Canagliflozin use in combination with phentermine resulted in additional weight loss compared to single agent monotherapy [133]. Similarly, SGLT-2 inhibitors combined with a GLP-1 agonist caused a greater weight reduction than the individual administration of each agent [134].

Metformin

Metformin was found to halt the weight gain caused by other anti-diabetic medications like insulin, thiazolidinediones, and sulfonylureas, or other medications associated with weight gains such as antipsychotics and antiretroviral therapy [135 - 139]. The weight changes are related to impaired glucose tolerance [140]. At the same time, metformin failed to reduce weight in obese people who do not have DM [141]. Possible mechanisms of metformin's effect on weight have been attributed to a reduction in the carbohydrate uptake in the gut, modulation of the vicious cycle of insulin resistance, reduction of lipid levels, and augmentation of the GLP-1 effects on fat cells [142 - 144].

ANTI-OBESITY MEDICATIONS

Orlistat

Orlistat received FDA approval in 1999. It inhibits gastric and pancreatic lipases, which leads to an approximately 30% decrease in the absorption of dietary fats. Due to such a mechanism, it is expected to have little effect on weight loss with non-fatty food consumption. The recommended dose is 120 mg thrice daily. It is contraindicated in pregnancy. No dose adjustment is needed in patients with kidney or hepatic impairment. The most common AEs are fatty/oily stool, flatus, fecal incontinence, increased defecation, abdominal cramps, and interfering with the absorption of fat-soluble vitamins [145]. Gastrointestinal AEs can be minimized by adding fiber-rich Psyllium [146]. Weight loss with orlistat is modest. Excess weight loss compared with placebo at 1 year is 2.6 kg (95% CI, -3.04 to -2.16 kg) [112].

Phentermine/topiramate

The FDA approved it in 2012 as the first combination agent for the long-term management of obesity. As this drug combination contains phentermine, it is a controlled DEA schedule IV substance. However, the European Medicines Agency (EMA) did not approve the medication due to concerns about abuse potential, psychiatric and cognitive AEs of topiramate, and the lack of long-term data on the cardiovascular effects of phentermine [147]. Phentermine-topiramate showed significant excess weight loss compared to placebo at 1 year of 8.8 kg (95% CI, -8.0 to -9.6 kg). A network meta-analysis also suggested that phentermine-topiramate, 15 mg/92 mg once daily, was associated with significant excess weight loss compared with other active agents, however, semaglutide was not included in that study (change *vs.* orlistat, 6.2 kg; *vs.* lorcaserin, 5.6 kg; *vs.* naltrexone-bupropion, 3.9 kg; and *vs.* liraglutide, 3.5 kg) [112].

The effect of phentermine, a sympathomimetic amine that is related to amphetamines, on weight management is likely mediated by the release of catecholamines in the hypothalamus, resulting in reduced appetite and decreased food consumption. The exact mechanism of action of topiramate on chronic weight management is unknown. The initial recommended dose is 3.75 mg/23 mg (phentermine 3.75 mg/topiramate 23 mg extended release) daily for 14 days; then increase to 7.5 mg/46 mg daily, to be administered once daily in the morning, as administration in the evening may cause insomnia. If 3% weight loss is not achieved after 12 weeks on a 7.5 mg/46 mg dose, the dose may be escalated, or use may be discontinued, and if 5% weight loss is not achieved after 12 weeks on a maximum daily dose of 15 mg/92 mg, use should be discontinued. Use should be discontinued gradually to prevent possible seizures. The maximum dose in patients with moderate or severe kidney impairment or patients with moderate hepatic impairment is 7.5 mg/46 mg.

Adverse effects include paresthesia, cognitive dysfunction, suicidal ideation, dry mouth, headache, insomnia, dysgeusia, constipation, tachycardia, and hypertension. Use should be avoided in patients with hypertension, arrhythmias, or cardiovascular diseases. Topiramate inhibits carbonic anhydrase and may cause the development of metabolic acidosis and nephrolithiasis. Due to the risk of oral clefts with topiramate, advice on contraceptive planning is imperative before this medication is prescribed to women of childbearing age [148].

Naltrexone/bupropion

A fixed-dose combination was approved by the FDA and the EMA in 2014. Bupropion, an antidepressant, inhibits dopamine and norepinephrine reuptake. Whereas naltrexone, a competitive opioid receptor antagonist approved for the treatment of opioid and alcohol addiction, antagonizes the opioid-dependent feedback loop that limits the effects of bupropion on the proopiomelanocortin neurons; hence, this drug combination works synergistically [149].

Side effects include nausea, constipation, headache, vomiting, dizziness, insomnia, dry mouth, and diarrhea. To minimize the side effects, the dose of naltrexone/bupropion should be escalated gradually with a starting dose of one combination tablet (8/90 mg) every morning for 1 week, followed by one tablet twice daily (BID) in the morning and evening at week 2, two tablets in the morning and one tablet in the evening at week 3, and two tablets BID (the maximum dose) at week 4. Patients should be monitored for depression or suicidal thoughts. Angle-closure glaucoma occurred in patients with untreated anatomically narrow angles treated with antidepressants. Concomitant treatment with CYP2B6 inhibitors (*e.g.*, ticlopidine or clopidogrel) can increase bupropion exposure. The maximum dose should be limited to one tablet twice daily when taken with CYP2B6 inhibitors. Contraindications are uncontrolled hypertension, seizure disorders, anorexia nervosa or bulimia, undergoing abrupt discontinuation of alcohol, benzodiazepines, barbiturates, and antiepileptic drugs, use of other bupropion-containing products, chronic opioid use, or use during or within 14 days of taking monoamine oxidase inhibitors [150]. Naltrexone-bupropion was associated with a significant excess of weight loss of 5.0 kg (95%CI, −5.94 to

−3.96 kg) compared to placebo at 1 year [112].

Surgical Options to Control Diabesity and Kidney Disease

Bariatric surgery (BS) is currently one of the most efficient treatments for obesity, with many beneficial long-term effects [151 - 153]. The basis for bariatric surgery benefit is the premise that obesity is a disease associated with adverse health effects which can be reversed after weight loss [154].

Prospective studies have demonstrated a beneficial effect of BS in controlling hypertension and DM in obese subjects [152, 153, 155]. In a systematic review and meta-analysis of 14 studies which included 29,208 patients who underwent BS, with a follow-up duration ranging from 2-14.7 years, weight loss among studies varied from 15-30% or more, and overall mortality was reduced by more than 50% (OR 0.48) [156].

According to the NIH consensus panel, bariatric surgery is indicated for patients with BMI >40 kg/m^2 and patients with BMI 35-40 with related comorbid conditions [154]. Obesity-related comorbidities are conditions either directly caused by overweight/obesity or known to contribute to the presence or severity of the condition and are expected to improve or remit with sustained weight loss [154].

Vertical sleeve gastrectomy (VSG) and Roux-en-Y gastric bypass (YGBR) are the most common procedures performed [87]. In YGBR, the stomach is transected, creating a gastric pouch with a nearly one-ounce capacity. A Roux-en-Y gastrojejunostomy is created, diverting nutrients from the stomach, duodenum, and proximal jejunum. VSG consists of resection of approximately 80% of the body of the stomach. Another less frequent procedure, biliopancreatic diversion with duodenal switch, consists of a sleeve gastrectomy anastomosed between the proximal duodenum and a bypassed intestine is created, leading to a degree of malabsorption.

Several non-surgical procedures have also been approved for obesity. In one of them, an adjustable gastric banding is placed in the proximal stomach, thus, the size of the gastric pouch and outlet is constricted. An intermittent vagal blockade is performed to reduce appetite and generate early satiety. Leads are placed about the vagal trunks at the diaphragm. Gastrointestinal endoscopic devices have also been used, such as a gastric balloon.

Complications include those related to the procedure, wound infection, dehydration, intestinal obstruction, marginal ulcer, ventral hernia, and metabolic complications such as vitamin deficiencies and weight regain. Thus, routine supplementation with vitamins, iron, minerals, calcium, and Vitamin D is recommended [154].

Emerging data suggest that T2DM is better controlled after RYGB than after VSG. The glucose-lowering effects of surgery depend on decreased peripheral insulin resistance, which depends on weight loss, and other mechanisms independent of weight loss [157 - 159]. The impact of BS on kidney function has been the focus of a variety of studies. In patients with CKD, weight loss after bariatric surgery has been associated with favorable effects, mainly through a reduction in proteinuria and stabilization of eGFR [87].

In the Teen-Longitudinal Assessment of Bariatric Surgery cohort [160], changes in kidney parameters were studied in 242 severely obese adolescents undergoing bariatric surgery. 17% of patients had albuminuria, and 3% had significantly decreased eGFR. After three years of follow-up in those with decreased kidney function at baseline, mean eGFR significantly improved from 76 to 102 ml/min/1.73 m2. Participants with albuminuria showed a significant improvement after surgery.

Data from the National Diabetes Register and the Scandinavian Obesity Surgery Register [152] showed small reductions in creatinine and albuminuria and stable eGFR among 5,321 individuals with T2DM who had undergone gastric bypass surgery. Patients in the gastric bypass group showed a lower incidence of albuminuria, with a 45% total risk reduction. There was also a lower incidence of CV disease and mortality (HR 0.36; 95% CI 0.7-0.97). Interestingly, these outcomes were consistent in all eGFR stages, and there was a 45% total risk reduction in albuminuria in the gastric bypass group. The composite of severe kidney disease or 50% reduction in eGFR was 44% lower after GBP (0.56; 95% CI 0.44-0.71)). Researchers concluded that large weight loss is more strongly associated with lowering cardiovascular and kidney complications.

In a meta-analysis that evaluated changes in kidney parameters after BS, authors reported the pooled risk ratio relative to baseline was reduced for both albuminuria (RR 0.42; 95% CI 0.36-0.50) and proteinuria (RR 0.31; 95% CI 0.22-0.43) [161]. Patients with hyperfiltration showed significant decreases in measured GFR, creatinine clearance, unadjusted eGFR, and adjusted eGFR. On the other hand, patients with baseline stage 2 CKD, showed a significant increase in eGFR after surgery.

The reduction of albuminuria may be related to weight-independent factors such as elevated GLP1 and/or restoration of vascular tone due to improved leptin and adiponectin levels, or procedures-specific reductions in insulin resistance [87]. In one trial, patients with CKD stages G1-3, A2,3, were randomly assigned to receive the best medical treatment *vs.* RYGB. At a 2-year follow-up, microalbuminuria remission occurred in 56.6% (95% CI 42-72) of patients in the medical treatment group and 81.6% (95% CI 70-94) of patients after RYGB. Therefore, bariatric surgery can help patients with CKD in 2 ways; one is to help delay the progression of CKD by reducing the glomerular damage caused by glomerular hyperfiltration. Second, in advanced CKD, it can increase patients' accessibility to kidney transplantation.

However, mechanisms implicated in the benefit of BS in kidney disease are not completely elucidated. Several studies evaluated RYGB surgery in experimental models. In the Zucker Diabetic Fatty (ZDF) rat model with obesity and DKD undergoing RYGB or diet restriction, RYGB induced weight loss and decreased urinary protein loss more effectively than diet restriction. On the other hand, urinary albumin excretion rate continued to increase over 8 weeks following duodenojejunal bypass (DJB) surgery, despite metabolic improvements in the context of ongoing weight gain [162, 163].

For patients with ESKD, the data about the safety and efficacy of these procedures in improving health outcomes is still unknown. Theoretically, BS can effectively increase patients' likelihood of kidney transplantation, as a high body mass index limits the patients' accessibility to transplantation. However, long-term data about the durability of weight loss from BS in ESKD patients, and whether it really affects patients' access to kidney transplantation and may improve graft survival is still lacking. The data about BS safety in ESKD patients are conflicting. One study compared the 30-day postoperative outcomes between dialysis-dependent and non-dialysis-dependent patients who underwent primary bariatric surgery between 2005 and 2013 in the USA. The study found no significant differences in the rate of adverse health outcomes [164]. Another study found a higher rate of complications with more advanced CKD stages, however, the absolute risk of complications was relatively low [165]. A recent study showed rapid growth in the use of BS in obese patients with ESKD in the US, which was driven largely by increases in the use of sleeve gastrectomy with low postoperative complications [166]. This increasing trend will probably provide more data that will inform clinical practice in the near future.

Both RYGB and DJB reduced creatinine clearance in the high fat-streptozotocin model, which suggested a reduction in glomerular hyperfiltration. Also, RYGB and matched weight-loss interventions showed equivalent reductions in glomerulomegaly and glomerulosclerosis [167]. DJB also reduced mesangial matrix expansion in the high-fat-streptozotocin model. Analyzing kidney transcriptomic response, RYGB decreased fibrosis, inflammation, and biological oxidation pathways at the transcriptomic level. No RYGB reversed the disease-associated increase in expression of TGFβ superfamily genes which correlated strongly with improvements in glomerular structure, and in the high-fat-streptozotocin model, Tgfb1 expression was reduced in glomerular and tubular epithelial cells 8 weeks post-RYGB, which suggests that bypass procedures might have an anti-fibrotic effect [168].

In humans, Morales *et al.* reported a decrease in adipokine, leptin, and visfatin levels among 12 patients after BS. Pro-inflammatory cytokines including IL-1B, IL-1ra, TNF-alpha, IL-6, and monocyte chemoattractant protein-1 (MCP-1) also showed a significant decrease, as well as VEGF and pro-fibrotic factors TGF-1 and TGF- β2 1 year after the procedure. A positive correlation was observed between the fold change of mentioned inflammatory parameters and proteinuria, in addition to a reduction in the NAFLD score of the subjects [169].

In conclusion, bariatric surgery is reported to have favorable kidney outcomes in patients with diabesity and should be considered as part of the treatment if indicated. Benefits seen in kidney and cardiovascular parameters might not be only related to weight loss; reduction of inflammatory markers seems to play a key role as well.

Future Pipeline Treatment

The currently approved anti-obesity medications still don't meet full patients' acceptance either in efficacy or safety. Despite Glucagon-Like Peptide 1 Receptor Agonists (GLP1-RA) becoming more and more accepted and approved as a treatment for obesity, scientists keep exploring new therapeutic options for the treatment of obesity.

As obesity arises from either increased caloric intake or decreased energy expenditure [170], understanding this mechanism helps in exploring new treatment options for obesity. New anti-obesity drugs can be classified into centrally acting, drugs with incretin-based effects, and others [171].

Centrally acting drugs act on central receptors causing either increased energy expenditure or anorexia and decreased intake. Setmelanotide is a synthetic melanocortin receptor 4 (MC4R) agonist, which decreases body weight and increases resting energy expenditure in human and animal trials [172]. Other centrally acting drugs, including cannabinoid type-1 receptor blockers, are still under the animal trials phase [173].

Tirzepatide

Tirzepatide (LY3298176) is a novel dual GIP and GLP-1 receptor agonist that is being developed for the treatment of T2DM. It has been designed to closely mimic the physiological incretin balance of GIP and GLP-1, as well as to prolong action by avoiding DPP-4 degradation [174].

In a double-blind, randomized, phase 2 study published in 2018, patients with T2D were randomly assigned to receive either once-weekly subcutaneous tirzepatide (1 mg, 5 mg, 10 mg, or 15 mg), dulaglutide (1.5 mg), or placebo for 26 weeks. More patients treated with 5 mg, 10 mg, and 15 mg tirzepatide reached weight targets (≥5%, ≥10%, and ≥15% weight loss from baseline) than those treated with placebo and dulaglutide. Changes in mean body weight ranged from −0.9 kg to −11.3 kg for tirzepatide (vs. −0.4 kg for placebo, −2.7 kg for dulaglutide). Changes in waist circumference ranged from −2.1 cm to −10.2 cm for tirzepatide (vs. −1.3 cm for placebo, −2.5 cm for dulaglutide). The incidence of gastrointestinal AEs with tirzepatide was dose-related (23.1% for 1 mg, 32.7% for 5 mg, 51.0% for 10 mg, and 66.0% for 15 mg tirzepatide, 42.6% for dulaglutide, 9.8% for placebo); most events were mild to moderate in intensity and transient. Decreased appetite was the second most common adverse event with tirzepatide (3.8% for 1 mg, 20.0% for 5 mg, 25.5% for 10 mg, 18.9% for 15 mg tirzepatide, 5.6% for dulaglutide, 2.0% for placebo). There were no reports of severe hypoglycemia [175].

The SURPASS trials studied the efficacy and safety of tirzepatide in adults with T2DM. The SURPASS-1 trial compared tirzepatide with a placebo. Tirzepatide was administered as a once-weekly subcutaneous injection for 40-weeks. Tirzepatide monotherapy induced a dose-dependent bodyweight loss ranging from 7.0 to 9.5 kg from a baseline of a body-mass index of 31.9 kg/m^2. The most frequent adverse events with tirzepatide were mild to moderate and transient gastrointestinal events, including nausea (12-18% *vs.* 6%), diarrhea (12-14% *vs.* 8%), and vomiting (2-6% *vs.* 2%) [176]. The SURPASS-2 trial compared tirzepatide at a dose of 5 mg, 10 mg, or 15 mg *versus* semaglutide at a dose of 1 mg. Tirzepatide at all doses was superior to semaglutide in reducing body weight (least-squares mean estimated treatment difference, -1.9 kg, -3.6 kg, and -5.5 kg, respectively; P<0.001 for all comparisons) [177]. The SURPASS-3 trial compared a once-weekly subcutaneous injection of tirzepatide (5, 10, or 15 mg) to a once-daily subcutaneous injection of titrated insulin degludec when added to metformin with or without SGLT2 inhibitors. Tirzepatide reduced mean body weight from baseline at week 52 by –7.5 kg to –12.9 kg (8.1–13.9% of baseline body weight), while insulin degludec caused a weight gain of 2.3 kg. Bodyweight reduction was observed for all doses of tirzepatide as early as 4 weeks after treatment initiation, and this continued until week 52 without reaching a plateau for any of the doses, irrespective of the occurrence of gastrointestinal adverse events [178]. The SURPASS-5 trial compared once-weekly subcutaneous injection of tirzepatide (5,

10, or 15 mg) *vs.* placebo, as an add-on to titrated insulin glargine with or without metformin. Bodyweight was reduced by tirzepatide (model estimate ± SD, -6.2 ± 0.58 kg, -8.2 ± 0.58 kg, and -10.9 ± 0.58 kg respectively for tirzepatide 5, 10, and 15 mg), while insulin glargine increased weight by 1.7 kg [179]. In all these trials, tirzepatide showed marked reductions in HbA1c and body weight when compared to other agents, including GLP-1 receptor agonists and insulin.

Cotadutide

Cotadutide (MEDI0382), is an experimental dual GLP-1 and glucagon receptor dual agonist. In published phase 2 trials, it showed a potential to reduce blood glucose and body weight in obese or overweight individuals with T2DM. In a randomized, placebo-controlled, double-blind, combined multiple-ascending dose trial, the body weight in the Cotadutide group was 3.84 kg (90% CI, − 4.55 − 3.12 kg) compared to 1.70 kg (90% CI, − 2.4 − 1.1 kg) in the placebo group (P = 0.0008). Gastrointestinal disorders (18 [72%] *vs.* 13 [40%]) and decreased appetite (five [20%] *vs.* none) occurred more frequently with MEDI0382 than placebo [180]. In another randomized, double-blind, phase 2a study, Cotadutide caused significant body weight reductions compared to placebo (-3.41% [-4.37, -2.44] *vs.* -0.08% [-1.45, 1.28]; P = 0.002). A significant increase in insulin area under the curve from 0 to 4 hours post-mixed-meal tolerance test was observed with Cotadutide and gastric emptying time was prolonged, suggesting actions mediated by enhanced insulin secretion and delayed gastric emptying [181].

Amylin Analogs

Amylin has glucoregulatory effects due to its co-secretion with insulin, induces satiety, suppresses postprandial glucagon secretion, and delays gastric emptying [182]. Pramlintide was FDA approved in 2005 for the treatment of type 1 and type 2 diabetes as an adjunct to insulin. In a systematic review and meta-analysis, pramlintide significantly reduced weight in patients with T2DM (−2.57 kg, [95% CI −3.44, −1.70], p < 0.00001) *versus* the control group. In obese patients without DM, pramlintide also caused a reduction in weight (−2.27 kg [95% CI −2.88, −1.66], p < 0.00001) [183].

Cagrilintide is a long-acting amylin analog. The combination of semaglutide with cagrilintide (each with dose escalation to 2.4 mg by subcutaneous injection once weekly) was recently studied. At week 20, mean percentage bodyweight reductions were greater with cagrilintide 1.2 and 2.4 mg than with placebo (15.7% [SE 1.6] for cagrilintide 1.2 mg and 17.1% [1.5] for cagrilintide 2.4 mg *vs.* 9.8% [1.2] for pooled placebo cohorts), and with cagrilintide 4.5 mg than with matched placebo (15.4% [1.3] *vs.* 8.0% [2.2]) [184].

Amylin mimetics (Davalinitide), a dual amylin and calcitonin receptor agonist, reduces food intake and body weight in animal trials [185, 186]. Leptin analogs had also been tried. Metreleptin, a human recombinant leptin injectable analog, reduced weight and is approved in Japan for lipodystrophic disorders [187, 188].

Leucine/Metformin/Sildenafil Combination

Mammalian sirtuin 1 (SIRT1) and AMP-activated protein kinase (AMPK) regulate lipid and energy metabolism, as they inhibit fat accumulation and stimulate fatty acid oxidation [189 - 191].

L-leucine is an activator of SIRT1, and metformin is a synergistic coactivator of sirtuin pathway signaling [192, 193]. SIRT1 also is stimulated by endothelial nitric oxide synthase (eNOS). Sildenafil, a phosphodiesterase 5 inhibitor, activate eNOS and increases nitric oxide production [194, 195]. These synergistic effects were demonstrated in the phase 2 randomized controlled trial. A fixed-dose combination of L-leucine 1.1 g, metformin 0.5 g, and sildenafil 1.0 mg for 16 weeks significantly reduced body weight (58% exhibited > 2-kg weight reduction *versus* 23% of placebo ($P = 0.027$), and 42% exhibited > 3-kg reduction in 16 weeks *versus* 0% for placebo ($P = 0.003$) [196].

Other novel targets include Beloranib, a Fumagilin analog that can cause robust weight loss and hypophagia in rat models, however, it causes sleep disorders and gastrointestinal side effects [197, 198]. Cetilistat, a pancreatic and gastric lipase inhibitor, causes weight loss and is better tolerated than orlistat [199]. Other proof of concept studies tested the effect of anti-obesity vaccines such as anti-Ghrelin and Somatostatin antibodies, which reduced food intake in animal studies [200]. However, no weight loss was shown in human clinical trials of the vaccine [201].

CONCLUSION

Diabesity has emerged as a major public health problem. Due to complex pathophysiology, patient-related factors, availability, costs, contraindications, and safety measures, pharmacological interventions should be individually tailored. Clinicians should choose the appropriate combinations of different anti-diabetic and/or anti-obesity drugs that achieve the patient's goals. Patients should be followed up closely to ensure the safety and efficacy of the treatment.

CONSENT FOR PUBLICATION

Not applicable.

CONFLICT OF INTEREST

The author declares no conflict of interest, financial or otherwise.

ACKNOWLEDGEMENTS

Declared none.

REFERENCES

[1] Ng ACT, Delgado V, Borlaug BA, Bax JJ. Diabesity: the combined burden of obesity and diabetes on heart disease and the role of imaging. Nat Rev Cardiol 2021; 18(4): 291-304. [http://dx.doi.org/10.1038/s41569-020-00465-5] [PMID: 33188304]

[2] Ogurtsova K, da Rocha Fernandes JD, Huang Y, *et al.* IDF Diabetes Atlas: Global estimates for the prevalence of diabetes for 2015 and 2040. Diabetes Res Clin Pract 2017; 128: 40-50. [http://dx.doi.org/10.1016/j.diabres.2017.03.024] [PMID: 28437734]

[3] Farag YMK, Gaballa MR. Diabesity: an overview of a rising epidemic. Nephrol Dial Transplant 2011; 26(1): 28-35. [http://dx.doi.org/10.1093/ndt/gfq576] [PMID: 21045078]

[4] Kalra S. Diabesity. J Pak Med Assoc 2013; 63(4): 532-4. [PMID: 23905459]

[5] Fu H, Liu S, Bastacky SI, Wang X, Tian XJ, Zhou D. Diabetic kidney diseases revisited: A new perspective for a new era. Mol Metab 2019; 30: 250-63. [http://dx.doi.org/10.1016/j.molmet.2019.10.005] [PMID: 31767176]

[6] KDOQI Clinical Practice Guidelines and Clinical Practice Recommendations for Diabetes and Chronic Kidney Disease. Am J Kidney Dis 2007; 49(2) (Suppl. 2).

[7] Fava S, Hattersley AT. The role of genetic susceptibility in diabetic nephropathy: evidence from family studies. Nephrol Dial Transplant 2002; 17(9): 1543-6. [http://dx.doi.org/10.1093/ndt/17.9.1543] [PMID: 12198200]

[8] Zatterale F, Longo M, Naderi J, Raciti GA, Desiderio A, Miele C, *et al.* Chronic Adipose Tissue Inflammation Linking Obesity to Insulin Resistance and Type 2 Diabetes. Frontiers in Physiology. Front Physiol 2020; Vol. 10.

[9] Tsuboi N, Okabayashi Y, Shimizu A, Yokoo T. The Renal Pathology of Obesity. Kidney Int Rep 2017; 2(2): 251-60. [http://dx.doi.org/10.1016/j.ekir.2017.01.007] [PMID: 29142961]

[10] Anders HJ, Huber TB, Isermann B, Schiffer M. CKD in diabetes: diabetic kidney disease *versus* nondiabetic kidney disease. Nat Rev Nephrol 2018; 14(6): 361-77. [http://dx.doi.org/10.1038/s41581-018-0001-y] [PMID: 29654297]

[11] Zagkotsis G, Markou M, Paschou E, Papanikolaou P, Sabanis N. Preventing the development and progression of diabetic kidney disease: Where do we stand? Diabetes Metab Syndr 2018; 12(4): 585-90. [http://dx.doi.org/10.1016/j.dsx.2018.03.012] [PMID: 29602760]

[12] Alicic RZ, Rooney MT, Tuttle KR. Diabetic Kidney Disease: Challenges, Progress, and Possibilities. Clinical Journal of the American Society of Nephrology 2017; 12(18): CJN.11491116.

[13] KDOQI Clinical Practice Guideline for Diabetes and CKD: 2012 Update. Am J Kidney Dis 2012; 60(5): 850-86. [http://dx.doi.org/10.1053/j.ajkd.2012.07.005] [PMID: 23067652]

[14] Chen Y, Lee K, Ni Z, He JC. Diabetic Kidney Disease: Challenges, Advances, and Opportunities. Kidney Dis 2020; 6(4): 215-25.
[http://dx.doi.org/10.1159/000506634] [PMID: 32903946]

[15] Bello AK, Levin A, Tonelli M, *et al.* Assessment of global kidney health care status. JAMA 2017; 317(18): 1864-81.
[http://dx.doi.org/10.1001/jama.2017.4046] [PMID: 28430830]

[16] Saran R, Robinson B, Abbott KC, *et al.* US Renal Data System 2018 Annual Data Report: Epidemiology of Kidney Disease in the United States. Am J Kidney Dis 2019; 73(3) (Suppl. 1): A7-8.
[http://dx.doi.org/10.1053/j.ajkd.2019.01.001] [PMID: 30798791]

[17] Pham TT, Sim JJ, Kujubu DA, Liu ILA, Kumar VA. Prevalence of nondiabetic renal disease in diabetic patients. Am J Nephrol 2007; 27(3): 322-8.
[http://dx.doi.org/10.1159/000102598] [PMID: 17495429]

[18] Sharma SG, Bomback AS, Radhakrishnan J, *et al.* The modern spectrum of renal biopsy findings in patients with diabetes. Clin J Am Soc Nephrol 2013; 8(10): 1718-24.
[http://dx.doi.org/10.2215/CJN.02510213] [PMID: 23886566]

[19] Seaquist ER, Goetz FC, Rich S, Barbosa J. Familial clustering of diabetic kidney disease. Evidence for genetic susceptibility to diabetic nephropathy. N Engl J Med 1989; 320(18): 1161-5.
[http://dx.doi.org/10.1056/NEJM198905043201801] [PMID: 2710189]

[20] Florez JC. Genetics of Diabetic Kidney Disease. Semin Nephrol 2016; 36(6): 474-80.
[http://dx.doi.org/10.1016/j.semnephrol.2016.09.012] [PMID: 27987549]

[21] Freedman BI, Tuttle AB, Spray BJ. Familial predisposition to nephropathy in African-Americans with non-insulin-dependent diabetes mellitus. Am J Kidney Dis 1995; 25(5): 710-3.
[http://dx.doi.org/10.1016/0272-6386(95)90546-4] [PMID: 7747724]

[22] Spray BJ, Atassi NG, Tuttle AB, Freedman BI. Familial risk, age at onset, and cause of end-stage renal disease in white Americans. J Am Soc Nephrol 1995; 5(10): 1806-10.
[http://dx.doi.org/10.1681/ASN.V5101806] [PMID: 7787148]

[23] Faronato PP, Maioli M, Tonolo G, *et al.* Clustering of albumin excretion rate abnormalities in Caucasian patients with NIDDM. Diabetologia 1997; 40(7): 816-23.
[http://dx.doi.org/10.1007/s001250050754] [PMID: 9243103]

[24] Canani LH, Gerchman F, Gross JL. Familial clustering of diabetic nephropathy in Brazilian type 2 diabetic patients. Diabetes 1999; 48(4): 909-13.
[http://dx.doi.org/10.2337/diabetes.48.4.909] [PMID: 10102711]

[25] Vijay V, Snehalatha C, Shina K, Lalitha S, Ramachandran A. Familial aggregation of diabetic kidney disease in Type 2 diabetes in south India. Diabetes Res Clin Pract 1999; 43(3): 167-71.
[http://dx.doi.org/10.1016/S0168-8227(99)00007-8] [PMID: 10369425]

[26] Dreyer G, Hull S, Mathur R, Chesser A, Yaqoob MM. Progression of chronic kidney disease in a multi-ethnic community cohort of patients with diabetes mellitus. Diabet Med 2013; 30(8): 956-63.
[http://dx.doi.org/10.1111/dme.12197] [PMID: 23600455]

[27] Coresh J, Selvin E, Stevens LA, *et al.* Prevalence of chronic kidney disease in the United States. JAMA 2007; 298(17): 2038-47.
[http://dx.doi.org/10.1001/jama.298.17.2038] [PMID: 17986697]

[28] Umeukeje EM, Young BA. Genetics and ESKD Disparities in African Americans. Am J Kidney Dis 2019; 74(6): 811-21.
[http://dx.doi.org/10.1053/j.ajkd.2019.06.006] [PMID: 31606237]

[29] Shepard BD. Sex differences in diabetes and kidney disease: mechanisms and consequences. 2019; 317(2): F456-62.
[http://dx.doi.org/10.1152/ajprenal.00249.2019]

[30] Pettitt DJ, Saad MF, Bennett PH, Nelson RG, Knowler WC. Familial predisposition to renal disease in two generations of Pima Indians with Type 2 (non-insulin-dependent) diabetes mellitus. Diabetologia 1990; 33(7): 438-43.
[http://dx.doi.org/10.1007/BF00404096] [PMID: 2401399]

[31] Mueller PW, Rogus JJ, Cleary PA, *et al.* Genetics of Kidneys in Diabetes (GoKinD) study: a genetics collection available for identifying genetic susceptibility factors for diabetic nephropathy in type 1 diabetes. J Am Soc Nephrol 2006; 17(7): 1782-90.
[http://dx.doi.org/10.1681/ASN.2005080822] [PMID: 16775037]

[32] Igo RP Jr, Iyengar SK, Nicholas SB, *et al.* Genomewide linkage scan for diabetic renal failure and albuminuria: the FIND study. Am J Nephrol 2011; 33(5): 381-9.
[http://dx.doi.org/10.1159/000326763] [PMID: 21454968]

[33] Sandholm N, Salem RM, McKnight AJ, *et al.* New susceptibility loci associated with kidney disease in type 1 diabetes. PLoS Genet 2012; 8(9): e1002921.
[http://dx.doi.org/10.1371/journal.pgen.1002921] [PMID: 23028342]

[34] Kato M, Natarajan R. Epigenetics and epigenomics in diabetic kidney disease and metabolic memory. Nat Rev Nephrol 2019; 15(6): 327-45.
[http://dx.doi.org/10.1038/s41581-019-0135-6] [PMID: 30894700]

[35] Sheng X, Qiu C, Liu H, *et al.* Systematic integrated analysis of genetic and epigenetic variation in diabetic kidney disease. Proc Natl Acad Sci USA 2020; 117(46): 29013-24.
[http://dx.doi.org/10.1073/pnas.2005905117] [PMID: 33144501]

[36] Agarwala V, Flannick J, Sunyaev S, Altshuler D. Evaluating empirical bounds on complex disease genetic architecture. Nat Genet 2013; 45(12): 1418-27.
[http://dx.doi.org/10.1038/ng.2804] [PMID: 24141362]

[37] Karaderi T, Drong AW, Lindgren CM. Insights into the Genetic Susceptibility to Type 2 Diabetes from Genome-Wide Association Studies of Obesity-Related Traits. Current Diabetes Reports. Curr Diab Rep 2015; Vol. 15.

[38] Ingelsson E, McCarthy MI. Human Genetics of Obesity and Type 2 Diabetes Mellitus. Circ Genom Precis Med 2018; 11(6): e002090.
[http://dx.doi.org/10.1161/CIRCGEN.118.002090] [PMID: 29899044]

[39] Graff M, Scott RA, Justice AE, *et al.* Genome-wide physical activity interactions in adiposity — A meta-analysis of 200,452 adults. PLoS Genet 2017; 13(4): e1006528.
[http://dx.doi.org/10.1371/journal.pgen.1006528] [PMID: 28448500]

[40] Yang Y, Liu B, Xia W, *et al.* FTO genotype and type 2 diabetes mellitus: Spatial analysis and meta-analysis of 62 case-control studies from different regions. Genes (Basel) 2017; 8(2): 70.
[http://dx.doi.org/10.3390/genes8020070] [PMID: 28208657]

[41] Cole JB, Florez JC. Genetics of diabetes mellitus and diabetes complications. Nat Rev Nephrol 2020; 16(7): 377-90.
[http://dx.doi.org/10.1038/s41581-020-0278-5] [PMID: 32398868]

[42] Pinto-Sietsma SJ, Navis G, Janssen WMT, de Zeeuw D, Gans ROB, de Jong PE. A central body fat distribution is related to renal function impairment, even in lean subjects. Am J Kidney Dis 2003; 41(4): 733-41.
[http://dx.doi.org/10.1016/S0272-6386(03)00020-9] [PMID: 12666059]

[43] Kramer H, Luke A, Bidani A, Cao G, Cooper R, McGee D. Obesity and prevalent and incident CKD: The hypertension detection and follow-up program. Am J Kidney Dis 2005; 46(4): 587-94.
[http://dx.doi.org/10.1053/j.ajkd.2005.06.007] [PMID: 16183412]

[44] Foster MC, Hwang SJ, Larson MG, *et al.* Overweight, obesity, and the development of stage 3 CKD: the Framingham Heart Study. Am J Kidney Dis 2008; 52(1): 39-48.
[http://dx.doi.org/10.1053/j.ajkd.2008.03.003] [PMID: 18440684]

[45] Chang A, Van Horn L, Jacobs DR Jr, *et al.* Lifestyle-related factors, obesity, and incident microalbuminuria: the CARDIA (Coronary Artery Risk Development in Young Adults) study. Am J Kidney Dis 2013; 62(2): 267-75.
[http://dx.doi.org/10.1053/j.ajkd.2013.02.363] [PMID: 23601954]

[46] Kovesdy C, Furth S, Zoccali C. Obesity and kidney disease: Hidden consequences of the epidemic. Indian J Nephrol 2017; 27(2): 85-92.
[http://dx.doi.org/10.4103/ijn.IJN_61_17] [PMID: 28356657]

[47] Tsuboi N, Utsunomiya Y, Kanzaki G, *et al.* Low glomerular density with glomerulomegaly in obesity-related glomerulopathy. Clin J Am Soc Nephrol 2012; 7(5): 735-41.
[http://dx.doi.org/10.2215/CJN.07270711] [PMID: 22403274]

[48] Siener R, Glatz S, Nicolay C, Hesse A. The role of overweight and obesity in calcium oxalate stone formation. Obes Res 2004; 12(1): 106-13.
[http://dx.doi.org/10.1038/oby.2004.14] [PMID: 14742848]

[49] Sharma K. The link between obesity and albuminuria: adiponectin and podocyte dysfunction. Kidney Int 2009; 76(2): 145-8.
[http://dx.doi.org/10.1038/ki.2009.137] [PMID: 19404275]

[50] Wolf G, Ziyadeh FN. Leptin and renal fibrosis. Contrib Nephrol 2006; 151: 175-83.
[http://dx.doi.org/10.1159/000095328] [PMID: 16929141]

[51] Ellington AA, Malik AR, Klee GG, Turner ST, Rule AD, Mosley TH, *et al.* Association of plasma resistin with glomerular filtration rate and albuminuria in hypertensive adults. Hypertension. Hypertension 2007; pp. 708-14.

[52] Pommer W. Preventive Nephrology: The Role of Obesity in Different Stages of Chronic Kidney Disease. Kidney Dis 2018; 4(4): 199-204.
[http://dx.doi.org/10.1159/000490247] [PMID: 30574496]

[53] Bailey MA. 11β-Hydroxysteroid Dehydrogenases and Hypertension in the Metabolic Syndrome. Current Hypertension Reports. Curr Hypertens Rep 2017; Vol. 19.

[54] Gant CM, Minovic I, Binnenmars H, *et al.* Lower renal function is associated with derangement of 11-β hydroxysteroid dehydrogenase in type 2 diabetes. J Endocr Soc 2018; 2(7): 609-20.
[http://dx.doi.org/10.1210/js.2018-00088] [PMID: 29942925]

[55] Standeven KF, Hess K, Carter AM, *et al.* Neprilysin, obesity and the metabolic syndrome. Int J Obes 2011; 35(8): 1031-40.
[http://dx.doi.org/10.1038/ijo.2010.227] [PMID: 21042321]

[56] Hall JE, do Carmo JM, da Silva AA, Wang Z, Hall ME. Obesity, kidney dysfunction and hypertension: mechanistic links. Nat Rev Nephrol 2019; 15(6): 367-85.
[http://dx.doi.org/10.1038/s41581-019-0145-4] [PMID: 31015582]

[57] Amorim RG, Guedes GS, Vasconcelos SML, Santos JCF. Kidney disease in diabetes mellitus: Cross-linking between hyperglycemia, redox imbalance and inflammation. Vol. 112, Arquivos Brasileiros de Cardiologia. Arq Bras Cardiol 2019; 112(5): 577-87.
[http://dx.doi.org/10.5935/abc.20190077] [PMID: 31188964]

[58] Zhang PP, Ge YC, Li SJ, Xie HL, Li LS, Liu ZH. Renal biopsy in type 2 diabetes: Timing of complications and evaluating of safety in Chinese patients. Nephrology (Carlton) 2011; 16(1): 100-5.
[http://dx.doi.org/10.1111/j.1440-1797.2010.01369.x] [PMID: 21175985]

[59] Kanwar YS, Sun L, Xie P, Liu F, Chen S. A glimpse of various pathogenetic mechanisms of diabetic nephropathy. Annu Rev Pathol 2011; 6(1): 395-423.
[http://dx.doi.org/10.1146/annurev.pathol.4.110807.092150] [PMID: 21261520]

[60] Suarez MLG, Thomas DB, Barisoni L, Fornoni A. Diabetic nephropathy: Is it time yet for routine kidney biopsy? World J Diabetes 2013; 4(6): 245-55.

[http://dx.doi.org/10.4239/wjd.v4.i6.245] [PMID: 24379914]

[61] Raparia K, Usman I, Kanwar YS. Renal morphologic lesions reminiscent of diabetic nephropathy. Arch Pathol Lab Med 2013; 137(3): 351-9.
[http://dx.doi.org/10.5858/arpa.2012-0243-RA] [PMID: 23451746]

[62] Schwartz M, Lewis EJ, Leonard-Martin T, Lewis JB, Batlle D, Berl T, *et al.* Renal pathology patterns in type II diabetes mellitus: relationship with retinopathy. Nephrol Dial Transplant 1998; 13(10): 2547-52.
[http://dx.doi.org/10.1093/ndt/13.10.2547] [PMID: 9794557]

[63] Fioretto P, Mauer M. Histopathology of diabetic nephropathy. Semin Nephrol 2007; 27(2): 195-207.
[http://dx.doi.org/10.1016/j.semnephrol.2007.01.012] [PMID: 17418688]

[64] Tuttle KR. Back to the future: Glomerular hyperfiltration and the diabetic kidney. Vol. 66, Diabetes. Diabetes 2017; 66(1): 14-6.
[http://dx.doi.org/10.2337/dbi16-0056] [PMID: 27999101]

[65] Heerspink HJL, Perkins BA, Fitchett DH, Husain M, Cherney DZI. Sodium Glucose Cotransporter 2 Inhibitors in the Treatment of Diabetes Mellitus. Circulation 2016; 134(10): 752-72.
[http://dx.doi.org/10.1161/CIRCULATIONAHA.116.021887] [PMID: 27470878]

[66] Grabias BM, Konstantopoulos K. The physical basis of renal fibrosis: effects of altered hydrodynamic forces on kidney homeostasis. Am J Physiol Renal Physiol 2014; 306(5): F473-85.
[http://dx.doi.org/10.1152/ajprenal.00503.2013]

[67] Zhu Q, Scherer PE. Immunologic and endocrine functions of adipose tissue: implications for kidney disease. Nat Rev Nephrol 2018; 14(2): 105-20.
[http://dx.doi.org/10.1038/nrneph.2017.157] [PMID: 29199276]

[68] Chawla T, Sharma D, Singh A. Role of the renin angiotensin system in diabetic nephropathy. World J Diabetes 2010; 1(5): 141-5.
[http://dx.doi.org/10.4239/wjd.v1.i5.141] [PMID: 21537441]

[69] Samsu N. Diabetic Nephropathy: Challenges in Pathogenesis, Diagnosis, and Treatment. BioMed Research International. Biomed Res Int 2021; Vol. 2021.

[70] Wellen KE, Hotamisligil GS. Inflammation, stress, and diabetes. J Clin Invest 2005; 115(5): 1111-9.
[http://dx.doi.org/10.1172/JCI25102] [PMID: 15864338]

[71] Duni A, Liakopoulos V, Roumeliotis S, Peschos D, Dounousi E. Oxidative stress in the pathogenesis and evolution of chronic kidney disease: Untangling ariadne's thread. Int J Mol Sci 2019; 20(15): 3711.
[http://dx.doi.org/10.3390/ijms20153711]

[72] Lim A. Diabetic nephropathy – complications and treatment. Int J Nephrol Renovasc Dis 2014; 7: 361-81.
[http://dx.doi.org/10.2147/IJNRD.S40172] [PMID: 25342915]

[73] Furuta T, Saito T, Ootaka T, *et al.* The role of macrophages in diabetic glomerulosclerosis. Am J Kidney Dis 1993; 21(5): 480-5.
[http://dx.doi.org/10.1016/S0272-6386(12)80393-3] [PMID: 8488815]

[74] Lindblom R, Higgins G, Coughlan M, de Haan JB. Targeting mitochondria and reactive oxygen species-driven pathogenesis in diabetic nephropathy. Rev Diabet Stud 2015; 12(1-2): 134-56.
[http://dx.doi.org/10.1900/RDS.2015.12.134] [PMID: 26676666]

[75] Tucker PS, Scanlan AT, Dalbo VJ. Chronic kidney disease influences multiple systems: Describing the relationship between oxidative stress, inflammation, kidney damage, and concomitant disease. Oxidative Medicine and Cellular Longevity. Oxid Med Cell Longev 2015; Vol. 2015.

[76] Permyakova A, Gammal A, Hinden L, Weitman M, Weinstock M, Tam J. A Novel Indoline Derivative Ameliorates Diabesity-Induced Chronic Kidney Disease by Reducing Metabolic

Abnormalities. Front Endocrinol (Lausanne) 2020; 11: 91.https://pubmed.ncbi.nlm.nih.gov/32218769/ [http://dx.doi.org/10.3389/fendo.2020.00091] [PMID: 32218769]

[77] Ogden CL, Yanovski SZ, Carroll MD, Flegal KM. The epidemiology of obesity. Gastroenterology 2007; 132(6): 2087-102.
[http://dx.doi.org/10.1053/j.gastro.2007.03.052] [PMID: 17498505]

[78] Stemmer K, Perez-Tilve D, Ananthakrishnan G, *et al.* High-fat-diet-induced obesity causes an inflammatory and tumor-promoting microenvironment in the rat kidney. Dis Model Mech 2012; 5(5): dmm.009407.
[http://dx.doi.org/10.1242/dmm.009407] [PMID: 22422828]

[79] Lee W, Eom DW, Jung Y, *et al.* Dendrobium moniliforme attenuates high-fat diet-induced renal damage in mice through the regulation of lipid-induced oxidative stress. Am J Chin Med 2012; 40(6): 1217-28.
[http://dx.doi.org/10.1142/S0192415X12500905] [PMID: 23227793]

[80] Declèves AE, Sharma K. New pharmacological treatments for improving renal outcomes in diabetes. Nat Rev Nephrol 2010; 6(6): 371-80.
[http://dx.doi.org/10.1038/nrneph.2010.57] [PMID: 20440278]

[81] Lagranha CJ, Fiorino P, Casarini DE, Schaan BDA, Irigoyen MC. Molecular bases of diabetic nephropathy. Arq Bras Endocrinol Metabol 2007; 51(6): 901-12.
[http://dx.doi.org/10.1590/S0004-27302007000600003] [PMID: 17934656]

[82] Machado UF, Schaan BD, Seraphim PM. Transportadores de glicose na síndrome metabólica. Arq Bras Endocrinol Metabol 2006; 50(2): 177-89.
[http://dx.doi.org/10.1590/S0004-27302006000200004] [PMID: 16767284]

[83] Hinden L, Kogot-Levin A, Tam J, Leibowitz G. Pathogenesis of diabesity-induced kidney disease: role of kidney nutrient sensing. FEBS Journal. FEBS J 2021.

[84] Rahmoune H, Thompson PW, Ward JM, Smith CD, Hong G, Brown J. Glucose transporters in human renal proximal tubular cells isolated from the urine of patients with non-insulin-dependent diabetes. Diabetes 2005; 54(12): 3427-34.
[http://dx.doi.org/10.2337/diabetes.54.12.3427] [PMID: 16306358]

[85] Solini A, Rossi C, Mazzanti CM, Proietti A, Koepsell H, Ferrannini E. Sodium-glucose co-transporter (SGLT)2 and SGLT1 renal expression in patients with type 2 diabetes. Diabetes Obes Metab 2017; 19(9): 1289-94.
[http://dx.doi.org/10.1111/dom.12970] [PMID: 28419670]

[86] Norton L, Shannon CE, Fourcaudot M, *et al.* Sodium-glucose co-transporter (SGLT) and glucose transporter (GLUT) expression in the kidney of type 2 diabetic subjects. Diabetes Obes Metab 2017; 19(9): 1322-6.
[http://dx.doi.org/10.1111/dom.13003] [PMID: 28477418]

[87] Docherty NG, le Roux CW. Bariatric surgery for the treatment of chronic kidney disease in obesity and type 2 diabetes mellitus. Nat Rev Nephrol 2020; 16(12): 709-20.
[http://dx.doi.org/10.1038/s41581-020-0323-4] [PMID: 32778788]

[88] Vallon V, Docherty NG. Intestinal regulation of urinary sodium excretion and the pathophysiology of diabetic kidney disease: a focus on glucagon-like peptide 1 and dipeptidyl peptidase 4. Exp Physiol 2014; 99(9): 1140-5.
[http://dx.doi.org/10.1113/expphysiol.2014.078766] [PMID: 25085841]

[89] Lay AC, Hurcombe JA, Betin VMS, *et al.* Prolonged exposure of mouse and human podocytes to insulin induces insulin resistance through lysosomal and proteasomal degradation of the insulin receptor. Diabetologia 2017; 60(11): 2299-311.
[http://dx.doi.org/10.1007/s00125-017-4394-0] [PMID: 28852804]

[90] Freedland ES. Role of a critical visceral adipose tissue threshold (CVATT) in metabolic syndrome:

Implications for controlling dietary carbohydrates: A review. Nutr Metab (Lond) 2004; 1(1): 12. [http://dx.doi.org/10.1186/1743-7075-1-12]

[91] Khan RS, Bril F, Cusi K, Newsome PN. Modulation of Insulin Resistance in Nonalcoholic Fatty Liver Disease. Hepatology 2019; 70(2): hep.30429. [http://dx.doi.org/10.1002/hep.30429] [PMID: 30556145]

[92] Smith BW, Adams LA. Nonalcoholic fatty liver disease and diabetes mellitus: pathogenesis and treatment. Nat Rev Endocrinol 2011; 7(8): 456-65. [http://dx.doi.org/10.1038/nrendo.2011.72] [PMID: 21556019]

[93] Loomba R, Sanyal AJ. The global NAFLD epidemic. Nat Rev Gastroenterol Hepatol 2013; 10(11): 686-90. [http://dx.doi.org/10.1038/nrgastro.2013.171] [PMID: 24042449]

[94] Marcuccilli M, Chonchol M. NAFLD and chronic kidney disease. Vol. 17, International Journal of Molecular Sciences. Int J Mol Sci 2016; 17(4): 562. [http://dx.doi.org/10.3390/ijms17040562]

[95] Aubert L, Sandino J, Gutiérrez-Solís E, García-Martín F, Segura J, Porrini E, *et al.* Role of non-alcoholic fatty liver disease in the evolution of renal function in patients with diabetes mellitus. Nephrol Dial Transplant 2021. [PMID: 33983444]

[96] Byrne CD, Targher G. NAFLD as a driver of chronic kidney disease. J Hepatol 2020; 72(4): 785-801. [http://dx.doi.org/10.1016/j.jhep.2020.01.013] [PMID: 32059982]

[97] Pappachan JM, Viswanath AK. Medical Management of Diabesity: Do We Have Realistic Targets?. Current Diabetes Reports. Curr Diab Rep 2017; Vol. 17.

[98] Franz MJ, Boucher JL, Rutten-Ramos S, VanWormer JJ. Lifestyle weight-loss intervention outcomes in overweight and obese adults with type 2 diabetes: a systematic review and meta-analysis of randomized clinical trials. J Acad Nutr Diet 2015; 115(9): 1447-63. [http://dx.doi.org/10.1016/j.jand.2015.02.031] [PMID: 25935570]

[99] Leitner DR, Frühbeck G, Yumuk V, *et al.* Obesity and type 2 diabetes: Two diseases with a need for combined treatment strategies - EASO can lead the way. Obes Facts 2017; 10(5): 483-92. [http://dx.doi.org/10.1159/000480525] [PMID: 29020674]

[100] Piercy KL, Troiano RP, Ballard RM, *et al.* The physical activity guidelines for Americans. JAMA 2018; 320(19): 2020-8. [http://dx.doi.org/10.1001/jama.2018.14854] [PMID: 30418471]

[101] Shai I, Schwarzfuchs D, Henkin Y, *et al.* Weight loss with a low-carbohydrate, Mediterranean, or low-fat diet. N Engl J Med 2008; 359(3): 229-41. [http://dx.doi.org/10.1056/NEJMoa0708681] [PMID: 18635428]

[102] Larsen RN, Mann NJ, Maclean E, Shaw JE. The effect of high-protein, low-carbohydrate diets in the treatment of type 2 diabetes: a 12 month randomised controlled trial. Diabetologia 2011; 54(4): 731-40. [http://dx.doi.org/10.1007/s00125-010-2027-y] [PMID: 21246185]

[103] Dansinger ML, Gleason JA, Griffith JL, Selker HP, Schaefer EJ. Comparison of the Atkins, Ornish, Weight Watchers, and Zone diets for weight loss and heart disease risk reduction: a randomized trial. JAMA 2005; 293(1): 43-53. [http://dx.doi.org/10.1001/jama.293.1.43] [PMID: 15632335]

[104] Curry SJ, Krist AH, Owens DK, Barry MJ, Caughey AB, Davidson KW, *et al.* Behavioral weight loss interventions to prevent obesity-relatedmorbidity and mortality in adults US preventive services task force recommendation statement. Vol. 320, JAMA - Journal of the American Medical Association. JAMA 2018; 1163-71. [PMID: 30326502]

[105] Wadden TA, Butryn ML, Hong PS, Tsai AG. Behavioral treatment of obesity in patients encountered in primary care settings: a systematic review. JAMA 2014; 312(17): 1779-91. [http://dx.doi.org/10.1001/jama.2014.14173] [PMID: 25369490]

[106] ADA standards of diabetes care 2021. Vol. 44. Diabetes Care 2021; S21-S226.

[107] Onakpoya IJ, Heneghan CJ, Aronson JK. Post-marketing withdrawal of anti-obesity medicinal products because of adverse drug reactions: a systematic review. BMC Med 2016; 14(1): 191. [http://dx.doi.org/10.1186/s12916-016-0735-y] [PMID: 27894343]

[108] U.S Food and Drug Administration. FDA requests the withdrawal of the weight-loss drug Belviq, Belviq XR (lorcaserin) from the market | FDA. Fda 2020; 3-5.

[109] Elrick H, Stimmler L, Hlad CJ Jr, Arai Y. PLASMA INSULIN RESPONSE TO ORAL AND INTRAVENOUS GLUCOSE ADMINISTRATION. J Clin Endocrinol Metab 1964; 24(10): 1076-82. [http://dx.doi.org/10.1210/jcem-24-10-1076] [PMID: 14228531]

[110] Baggio LL, Drucker DJ. Biology of Incretins: GLP-1 and GIP. Gastroenterology 2007; 132(6): 2131-57. [http://dx.doi.org/10.1053/j.gastro.2007.03.054] [PMID: 17498508]

[111] Food and Drug Administration. VICTOZA® (liraglutide). PRESCRIBING INFORMATION 2010; p. 3.

[112] Khera R, Murad MH, Chandar AK, *et al.* Association of Pharmacological Treatments for Obesity With Weight Loss and Adverse Events. JAMA 2016; 315(22): 2424-34. [http://dx.doi.org/10.1001/jama.2016.7602] [PMID: 27299618]

[113] Monami M, Nreu B, Scatena A, *et al.* Safety issues with glucagon-like peptide-1 receptor agonists (pancreatitis, pancreatic cancer and cholelithiasis): D ata from randomized controlled trials. Diabetes Obes Metab 2017; 19(9): 1233-41. [http://dx.doi.org/10.1111/dom.12926] [PMID: 28244632]

[114] Marso SP, Daniels GH, Brown-Frandsen K, *et al.* Liraglutide and Cardiovascular Outcomes in Type 2 Diabetes. N Engl J Med 2016; 375(4): 311-22. [http://dx.doi.org/10.1056/NEJMoa1603827] [PMID: 27295427]

[115] Blonde L, Jendle J, Gross J, *et al.* Once-weekly dulaglutide *versus* bedtime insulin glargine, both in combination with prandial insulin lispro, in patients with type 2 diabetes (AWARD-4): a randomised, open-label, phase 3, non-inferiority study. Lancet 2015; 385(9982): 2057-66. [http://dx.doi.org/10.1016/S0140-6736(15)60936-9] [PMID: 26009229]

[116] Nauck M, Weinstock RS, Umpierrez GE, Guerci B, Skrivanek Z, Milicevic Z. Efficacy and safety of dulaglutide *versus* sitagliptin after 52 weeks in type 2 diabetes in a randomized controlled trial (AWARD-5). Diabetes Care 2014; 37(8): 2149-58. [http://dx.doi.org/10.2337/dc13-2761] [PMID: 24742660]

[117] Bonora E, Frias JP, Tinahones FJ, *et al.* Effect of dulaglutide 3.0 and 4.5 mg on weight in patients with type 2 diabetes: Exploratory analyses of AWARD -11. Diabetes Obes Metab 2021; 23(10): 2242-50. [http://dx.doi.org/10.1111/dom.14465] [PMID: 34189841]

[118] Wilding JPH, Batterham RL, Calanna S, *et al.* Once-Weekly Semaglutide in Adults with Overweight or Obesity. N Engl J Med 2021; 384(11): 989-1002. [http://dx.doi.org/10.1056/NEJMoa2032183] [PMID: 33567185]

[119] Davies M, Færch L, Jeppesen OK, *et al.* Semaglutide 2·4 mg once a week in adults with overweight or obesity, and type 2 diabetes (STEP 2): a randomised, double-blind, double-dummy, placebo-controlled, phase 3 trial. Lancet 2021; 397(10278): 971-84. [http://dx.doi.org/10.1016/S0140-6736(21)00213-0] [PMID: 33667417]

[120] Wadden TA, Bailey TS, Billings LK, *et al.* Effect of Subcutaneous Semaglutide *vs.* Placebo as an Adjunct to Intensive Behavioral Therapy on Body Weight in Adults With Overweight or Obesity.

JAMA 2021; 325(14): 1403-13.
[http://dx.doi.org/10.1001/jama.2021.1831] [PMID: 33625476]

[121] Rubino D, Abrahamsson N, Davies M, *et al.* Effect of Continued Weekly Subcutaneous Semaglutide *vs.* Placebo on Weight Loss Maintenance in Adults With Overweight or Obesity. JAMA 2021; 325(14): 1414-25.
[http://dx.doi.org/10.1001/jama.2021.3224] [PMID: 33755728]

[122] Viljoen A, Bluher M, Chow FCC, Roux CLE, Rosenstock J, Lausvig NL, *et al.* Semaglutide Reduces Body Weight *vs.* Dulaglutide across Baseline BMI Subgroups in SUSTAIN 7. Diabetes 2018; 67(Supplement 1): 1083.

[123] Capehorn MS, Catarig AM, Furberg JK, *et al.* Efficacy and safety of once-weekly semaglutide 1.0 mg *vs.* once-daily liraglutide 1.2 mg as add-on to 1–3 oral antidiabetic drugs in subjects with type 2 diabetes (SUSTAIN 10). Diabetes Metab 2020; 46(2): 100-9.
[http://dx.doi.org/10.1016/j.diabet.2019.101117] [PMID: 31539622]

[124] O'Neil PM, Birkenfeld AL, McGowan B, *et al.* Efficacy and safety of semaglutide compared with liraglutide and placebo for weight loss in patients with obesity: a randomised, double-blind, placebo and active controlled, dose-ranging, phase 2 trial. Lancet 2018; 392(10148): 637-49.
[http://dx.doi.org/10.1016/S0140-6736(18)31773-2] [PMID: 30122305]

[125] U.S Food and Drug Administration. Wegovy (semaglutide) product information. 2021.

[126] Marso SP, Bain SC, Consoli A, *et al.* Semaglutide and Cardiovascular Outcomes in Patients with Type 2 Diabetes. N Engl J Med 2016; 375(19): 1834-44.
[http://dx.doi.org/10.1056/NEJMoa1607141] [PMID: 27633186]

[127] Liu XY, Zhang N, Chen R, Zhao JG, Yu P. Efficacy and safety of sodium–glucose cotransporter 2 inhibitors in type 2 diabetes: a meta-analysis of randomized controlled trials for 1 to 2 years. J Diabetes Complications 2015; 29(8): 1295-303.
[http://dx.doi.org/10.1016/j.jdiacomp.2015.07.011] [PMID: 26365905]

[128] Bays HE, Weinstein R, Law G, Canovatchel W. Canagliflozin: Effects in overweight and obese subjects without diabetes mellitus. Obesity (Silver Spring) 2014; 22(4): 1042-9.
[http://dx.doi.org/10.1002/oby.20663] [PMID: 24227660]

[129] Chen M, Wang H, Cui W, Xu H, Zheng Q. Effect of SGLT inhibitors on weight and lipid metabolism at 24 weeks of treatment in patients with diabetes mellitus. Medicine (Baltimore) 2021; 100(6): e24593.
[http://dx.doi.org/10.1097/MD.0000000000024593] [PMID: 33578559]

[130] Rajeev SP, Cuthbertson DJ, Wilding JPH. Energy balance and metabolic changes with sodium-glucose co-transporter 2 inhibition. Diabetes Obes Metab 2016; 18(2): 125-34.
[http://dx.doi.org/10.1111/dom.12578] [PMID: 26403227]

[131] Brown E, Wilding JPH, Barber TM, Alam U, Cuthbertson DJ. Weight loss variability with SGLT2 inhibitors and GLP-1 receptor agonists in type 2 diabetes mellitus and obesity: Mechanistic possibilities. Obes Rev 2019; 20(6): 816-28.
[http://dx.doi.org/10.1111/obr.12841] [PMID: 30972878]

[132] Janež A, Fioretto P. SGLT2 Inhibitors and the Clinical Implications of Associated Weight Loss in Type 2 Diabetes: A Narrative Review. Diabetes Ther 2021; 12(8): 2249-61.
[http://dx.doi.org/10.1007/s13300-021-01104-z] [PMID: 34244976]

[133] Hollander P, Bays HE, Rosenstock J, *et al.* Coadministration of Canagliflozin and Phentermine for Weight Management in Overweight and Obese Individuals Without Diabetes: A Randomized Clinical Trial. Diabetes Care 2017; 40(5): 632-9.
[http://dx.doi.org/10.2337/dc16-2427] [PMID: 28289041]

[134] Frías JP, Guja C, Hardy E, *et al.* Exenatide once weekly plus dapagliflozin once daily *versus* exenatide or dapagliflozin alone in patients with type 2 diabetes inadequately controlled with metformin

monotherapy (DURATION-8): a 28 week, multicentre, double-blind, phase 3, randomised controlled trial. Lancet Diabetes Endocrinol 2016; 4(12): 1004-16.
[http://dx.doi.org/10.1016/S2213-8587(16)30267-4] [PMID: 27651331]

[135] Effect of intensive blood-glucose control with metformin on complications in overweight patients with type 2 diabetes (UKPDS 34). Lancet 1998; 352(9131): 854-65.
[http://dx.doi.org/10.1016/S0140-6736(98)07037-8] [PMID: 9742977]

[136] Kahn SE, Haffner SM, Heise MA, *et al.* Glycemic durability of rosiglitazone, metformin, or glyburide monotherapy. N Engl J Med 2006; 355(23): 2427-43.
[http://dx.doi.org/10.1056/NEJMoa066224] [PMID: 17145742]

[137] Lund SS, Tarnow L, Frandsen M, *et al.* Combining insulin with metformin or an insulin secretagogue in non-obese patients with type 2 diabetes: 12 month, randomised, double blind trial. BMJ 2009; 339(nov09 1): b4324.
[http://dx.doi.org/10.1136/bmj.b4324] [PMID: 19900993]

[138] Wu RR, Zhao JP, Jin H, *et al.* Lifestyle intervention and metformin for treatment of antipsychotic-induced weight gain: a randomized controlled trial. JAMA 2008; 299(2): 185-93.
[http://dx.doi.org/10.1001/jama.2007.56-b] [PMID: 18182600]

[139] Hadigan C, Corcoran C, Basgoz N, Davis B, Sax P, Grinspoon S. Metformin in the treatment of HIV lipodystrophy syndrome: A randomized controlled trial. JAMA 2000; 284(4): 472-7.
[http://dx.doi.org/10.1001/jama.284.4.472] [PMID: 10904511]

[140] Lily M, Godwin M. Treating prediabetes with metformin: systematic review and meta-analysis. Can Fam Physician 2009; 55(4): 363-9.
[PMID: 19366942]

[141] Desilets AR, Dhakal-Karki S, Dunican KC. Role of metformin for weight management in patients without type 2 diabetes. Ann Pharmacother 2008; 42(6): 817-26.
[http://dx.doi.org/10.1345/aph.1K656] [PMID: 18477733]

[142] Glueck CJ, Fontaine RN, Wang P, *et al.* Metformin reduces weight, centripetal obesity, insulin, leptin, and low-density lipoprotein cholesterol in nondiabetic, morbidly obese subjects with body mass index greater than 30. Metabolism 2001; 50(7): 856-61.
[http://dx.doi.org/10.1053/meta.2001.24192] [PMID: 11436194]

[143] Kusaka I, Nagasaka S, Horie H, Ishibashi S. Metformin, but not pioglitazone, decreases postchallenge plasma ghrelin levels in type 2 diabetic patients: a possible role in weight stability? Diabetes Obes Metab 2008; 10(11): 1039-46.
[http://dx.doi.org/10.1111/j.1463-1326.2008.00857.x] [PMID: 18355332]

[144] Mannucci E, Ognibene A, Cremasco F, *et al.* Effect of metformin on glucagon-like peptide 1 (GLP-1) and leptin levels in obese nondiabetic subjects. Diabetes Care 2001; 24(3): 489-94.
[http://dx.doi.org/10.2337/diacare.24.3.489] [PMID: 11289473]

[145] Xenical FDA. orlistat. PRESCRIBING INFORMATION 1999.

[146] Cavaliere H, Floriano I, Medeiros-Neto G. Gastrointestinal side effects of orlistat may be prevented by concomitant prescription of natural fibers (psyllium mucilloid). International journal of obesity and related metabolic disorders. Journal of the International Association for the Study of Obesity 2001; 25(7): 1095-9.

[147] Assessment report. Qsiva 2013; 44(February): 5-67.

[148] Food and Drug Administration. QSYMIA (phentermine and topiramate extended-release). PRESCRIBING INFORMATION 2012.

[149] Greenway FL, Whitehouse MJ, Guttadauria M, *et al.* Rational design of a combination medication for the treatment of obesity. Obesity (Silver Spring) 2009; 17(1): 30-9.
[http://dx.doi.org/10.1038/oby.2008.461] [PMID: 18997675]

[150] Food and Drug Administration. CONTRAVE® (naltrexone HCl and bupropion HCl). PRESCRIBING INFORMATION 2014.

[151] Inge TH, Courcoulas AP, Jenkins TM, *et al.* Weight Loss and Health Status 3 Years after Bariatric Surgery in Adolescents. N Engl J Med 2016; 374(2): 113-23. [http://dx.doi.org/10.1056/NEJMoa1506699] [PMID: 26544725]

[152] Liakopoulos V, Franzén S, Svensson AM, *et al.* Renal and cardiovascular outcomes after weight loss from gastric bypass surgery in type 2 diabetes: Cardiorenal risk reductions exceed atherosclerotic benefits. Diabetes Care 2020; 43(6): 1276-84. [http://dx.doi.org/10.2337/dc19-1703] [PMID: 32152136]

[153] Sjöström L. Review of the key results from the Swedish Obese Subjects (SOS) trial - a prospective controlled intervention study of bariatric surgery. J Intern Med 2013; 273(3): 219-34. [http://dx.doi.org/10.1111/joim.12012] [PMID: 23163728]

[154] Wolfe BM, Kvach E, Eckel RH. Treatment of Obesity. Circ Res 2016; 118(11): 1844-55. [http://dx.doi.org/10.1161/CIRCRESAHA.116.307591] [PMID: 27230645]

[155] Maric-Bilkan C. Obesity and diabetic kidney disease. Med Clin North Am 2013; 97(1): 59-74. [http://dx.doi.org/10.1016/j.mcna.2012.10.010] [PMID: 23290730]

[156] Schiavon CA, Bersch-Ferreira AC, Santucci EV, *et al.* Effects of bariatric surgery in obese patients with hypertension the GATEWAY randomized trial (gastric bypass to treat obese patients with steady hypertension). Circulation 2018; 137(11): 1132-42. [http://dx.doi.org/10.1161/CIRCULATIONAHA.117.032130] [PMID: 29133606]

[157] Quercia I, Dutia R, Kotler DP, Belsley S, Laferrère B. Gastrointestinal changes after bariatric surgery. Diabetes Metab 2014; 40(2): 87-94. [http://dx.doi.org/10.1016/j.diabet.2013.11.003] [PMID: 24359701]

[158] Roux CW, Aylwin SJB, Batterham RL, *et al.* Gut hormone profiles following bariatric surgery favor an anorectic state, facilitate weight loss, and improve metabolic parameters. Ann Surg 2006; 243(1): 108-14. [http://dx.doi.org/10.1097/01.sla.0000183349.16877.84] [PMID: 16371744]

[159] Bojsen-Møller KN, Dirksen C, Jørgensen NB, *et al.* Early enhancements of hepatic and later of peripheral insulin sensitivity combined with increased postprandial insulin secretion contribute to improved glycemic control after Roux-en-Y gastric bypass. Diabetes 2014; 63(5): 1725-37. [http://dx.doi.org/10.2337/db13-1307] [PMID: 24241533]

[160] Nehus EJ, Khoury JC, Inge TH, *et al.* Kidney outcomes three years after bariatric surgery in severely obese adolescents. Kidney Int 2017; 91(2): 451-8. [http://dx.doi.org/10.1016/j.kint.2016.09.031] [PMID: 27914704]

[161] Li K, Zou J, Ye Z, *et al.* Effects of bariatric surgery on renal function in obese patients: A systematic review and meta analysis. Vol. 11, PLoS ONE. PLoS One 2016; 11(10): e0163907. [http://dx.doi.org/10.1371/journal.pone.0163907]

[162] Zhiqing W, Jing W, Haili X, *et al.* Renal function is ameliorated in a diabetic nephropathy rat model through a duodenal-jejunal bypass. Diabetes Res Clin Pract 2014; 103(1): 26-34. [http://dx.doi.org/10.1016/j.diabres.2013.12.001] [PMID: 24398318]

[163] Wu D, Cheng Y, Huang X, Zhong M, Liu S, Hu S. Downregulation of lncRNA MALAT1 contributes to renal functional improvement after duodenal-jejunal bypass in a diabetic rat model. J Physiol Biochem 2018; 74(3): 431-9. [http://dx.doi.org/10.1007/s13105-018-0636-y] [PMID: 29781038]

[164] Andalib A, Aminian A, Khorgami Z, Navaneethan SD, Schauer PR, Brethauer SA. Safety analysis of primary bariatric surgery in patients on chronic dialysis. Surg Endosc 2016; 30(6): 2583-91.https://pubmed.ncbi.nlm.nih.gov/26416373/ [http://dx.doi.org/10.1007/s00464-015-4530-1] [PMID: 26416373]

[165] Turgeon NA, Perez S, Mondestin M, *et al.* The impact of renal function on outcomes of bariatric surgery. J Am Soc Nephrol 2012; 23(5): 885-94.https://pubmed.ncbi.nlm.nih.gov/22383694/ [http://dx.doi.org/10.1681/ASN.2011050476] [PMID: 22383694]

[166] Sheetz KH, Woodside KJ, Shahinian VB, Dimick JB, Montgomery JR, Waits SA. Trends in Bariatric Surgery Procedures among Patients with ESKD in the United States. Clin J Am Soc Nephrol 2019; 14(8): 1193-9.https://pubmed.ncbi.nlm.nih.gov/31345840/ [http://dx.doi.org/10.2215/CJN.01480219] [PMID: 31345840]

[167] Neff KJ, Elliott JA, Corteville C, *et al.* Effect of Roux-en-Y gastric bypass and diet-induced weight loss on diabetic kidney disease in the Zucker diabetic fatty rat. Surg Obes Relat Dis 2017; 13(1): 21-7. [http://dx.doi.org/10.1016/j.soard.2016.08.026] [PMID: 27665150]

[168] He B, Yu C, Du R, Wang Y, Han P. Roux-en-Y esophagojejunostomy reduces serum and aortic inflammatory biomarkers in type 2 diabetic rats. Obes Surg 2014; 24(6): 916-26. [http://dx.doi.org/10.1007/s11695-014-1195-0] [PMID: 24573962]

[169] Morales E, Porrini E, Martin-Taboada M, *et al.* Renoprotective role of bariatric surgery in patients with established chronic kidney disease. Clin Kidney J 2021; 14(9): 2037-46. [http://dx.doi.org/10.1093/ckj/sfaa266] [PMID: 34476090]

[170] Hill JO. Understanding and addressing the epidemic of obesity: an energy balance perspective. Endocr Rev 2006; 27(7): 750-61. [http://dx.doi.org/10.1210/er.2006-0032] [PMID: 17122359]

[171] Srivastava G, Apovian C. Future Pharmacotherapy for Obesity: New Anti-obesity Drugs on the Horizon. Curr Obes Rep 2018; 7(2): 147-61. [http://dx.doi.org/10.1007/s13679-018-0300-4] [PMID: 29504049]

[172] Fani L, Bak S, Delhanty P, Van Rossum EFC, Van Den Akker ELT. The melanocortin-4 receptor as target for obesity treatment: A systematic review of emerging pharmacological therapeutic options. International Journal of Obesity. Nature Publishing Group 2014; Vol. 38: pp. 163-9.

[173] Salamone JD, McLaughlin PJ, Sink K, Makriyannis A, Parker LA. Cannabinoid CB1 receptor inverse agonists and neutral antagonists: Effects on food intake, food-reinforced behavior and food aversions. Physiol Behav 2007; 91(4): 383-8. [http://dx.doi.org/10.1016/j.physbeh.2007.04.013] [PMID: 17521686]

[174] Bailey CJ. Tirzepatide: a new low for bodyweight and blood glucose. Lancet Diabetes Endocrinol 2021; 9(10): 646-8. [http://dx.doi.org/10.1016/S2213-8587(21)00217-5] [PMID: 34419226]

[175] Frias JP, Nauck MA, Van J, *et al.* Efficacy and safety of LY3298176, a novel dual GIP and GLP-1 receptor agonist, in patients with type 2 diabetes: a randomised, placebo-controlled and active comparator-controlled phase 2 trial. Lancet 2018; 392(10160): 2180-93. [http://dx.doi.org/10.1016/S0140-6736(18)32260-8] [PMID: 30293770]

[176] Rosenstock J, Wysham C, Frías JP, *et al.* Efficacy and safety of a novel dual GIP and GLP-1 receptor agonist tirzepatide in patients with type 2 diabetes (SURPASS-1): a double-blind, randomised, phase 3 trial. Lancet 2021; 398(10295): 143-55. [http://dx.doi.org/10.1016/S0140-6736(21)01324-6] [PMID: 34186022]

[177] Frías JP, Davies MJ, Rosenstock J, *et al.* Tirzepatide *versus* Semaglutide Once Weekly in Patients with Type 2 Diabetes. N Engl J Med 2021; 385(6): 503-15. [http://dx.doi.org/10.1056/NEJMoa2107519] [PMID: 34170647]

[178] Ludvik B, Giorgino F, Jódar E, *et al.* Once-weekly tirzepatide *versus* once-daily insulin degludec as add-on to metformin with or without SGLT2 inhibitors in patients with type 2 diabetes (SURPASS-3): a randomised, open-label, parallel-group, phase 3 trial. Lancet 2021; 398(10300): 583-98. [http://dx.doi.org/10.1016/S0140-6736(21)01443-4] [PMID: 34370970]

[179] Dahl D, Onishi Y, Norwood P, Huh R, Patel H, Rodríguez Á. 80-LB: Tirzepatide, a Dual GIP/GLP-1

Receptor Agonist, Is Effective and Safe When Added to Basal Insulin for Treatment of Type 2 Diabetes (SURPASS-5). Diabetes 2021; 70 (Suppl. 1): 80-LB. [-LB.]. [http://dx.doi.org/10.2337/db21-80-LB]

[180] Ambery P, Parker VE, Stumvoll M, *et al.* MEDI0382, a GLP-1 and glucagon receptor dual agonist, in obese or overweight patients with type 2 diabetes: a randomised, controlled, double-blind, ascending dose and phase 2a study. Lancet 2018; 391(10140): 2607-18. [http://dx.doi.org/10.1016/S0140-6736(18)30726-8] [PMID: 29945727]

[181] Parker VER, Robertson D, Wang T, *et al.* Efficacy, Safety, and Mechanistic Insights of Cotadutide, a Dual Receptor Glucagon-Like Peptide-1 and Glucagon Agonist. J Clin Endocrinol Metab 2020; 105(3): 803-20. Parker VER, [http://dx.doi.org/10.1210/clinem/dgz047] [PMID: 31608926]

[182] Mathiesen DS, Lund A, Vilsbøll T, Knop FK, Bagger JI. Amylin and Calcitonin: Potential Therapeutic Strategies to Reduce Body Weight and Liver Fat. Front Endocrinol (Lausanne) 2021; 11: 617400. [http://dx.doi.org/10.3389/fendo.2020.617400]

[183] Singh-Franco D, Perez A, Harrington C. The effect of pramlintide acetate on glycemic control and weight in patients with type 2 diabetes mellitus and in obese patients without diabetes: a systematic review and meta-analysis. Diabetes Obes Metab 2011; 13(2): 169-80. [http://dx.doi.org/10.1111/j.1463-1326.2010.01337.x] [PMID: 21199269]

[184] Enebo LB, Berthelsen KK, Kankam M, *et al.* Safety, tolerability, pharmacokinetics, and pharmacodynamics of concomitant administration of multiple doses of cagrilintide with semaglutide 2·4 mg for weight management: a randomised, controlled, phase 1b trial. Lancet 2021; 397(10286): 1736-48. [http://dx.doi.org/10.1016/S0140-6736(21)00845-X]

[185] Lutz TA. Control of food intake and energy expenditure by amylin—therapeutic implications. International Journal of Obesity 2009; 33(1): S24-7.

[186] Hjuler ST, Gydesen S, Andreassen KV, *et al.* The dual amylin- and calcitonin-receptor agonist KBP-042 increases insulin sensitivity and induces weight loss in rats with obesity. Obesity (Silver Spring) 2016; 24(8): 1712-22. [http://dx.doi.org/10.1002/oby.21563] [PMID. 27296301]

[187] Chou K, Perry CM. Metreleptin: First Global Approval. Drugs 2013; 73(9): 989-7.

[188] Tchang BG, Shukla AP, Aronne LJ. Metreleptin and generalized lipodystrophy and evolving therapeutic perspectives 2015; 15(7): 1061-75. [http://dx.doi.org/10.1517/14712598.2015.1052789]

[189] Cantó C, Gerhart-Hines Z, Feige JN, *et al.* AMPK regulates energy expenditure by modulating NAD+ metabolism and SIRT1 activity. Nature 2009; 458(7241): 1056-60. [http://dx.doi.org/10.1038/nature07813] [PMID: 19262508]

[190] Hou X, Xu S, Maitland-Toolan KA, *et al.* SIRT1 regulates hepatocyte lipid metabolism through activating AMP-activated protein kinase. J Biol Chem 2008; 283(29): 20015-26. [http://dx.doi.org/10.1074/jbc.M802187200] [PMID: 18482975]

[191] Ruderman NB, Carling D, Prentki M, Cacicedo JM. AMPK, insulin resistance, and the metabolic syndrome. J Clin Invest 2013; 123(7): 2764-72. [http://dx.doi.org/10.1172/JCI67227] [PMID: 23863634]

[192] Bruckbauer A, Zemel MB, Thorpe T, *et al.* Synergistic effects of leucine and resveratrol on insulin sensitivity and fat metabolism in adipocytes and mice. Nutr Metab (Lond) 2012; 9(1): 77. [http://dx.doi.org/10.1186/1743-7075-9-77] [PMID: 22913271]

[193] Fu L, Bruckbauer A, Li F, *et al.* Leucine amplifies the effects of metformin on insulin sensitivity and glycemic control in diet-induced obese mice. Metabolism 2015; 64(7): 845-56. [http://dx.doi.org/10.1016/j.metabol.2015.03.007] [PMID: 25858853]

[194] Musicki B, Bivalacqua TJ, Champion HC, Burnett AL. Sildenafil promotes eNOS activation and inhibits NADPH oxidase in the transgenic sickle cell mouse penis. J Sex Med 2014; 11(2): 424-30. [http://dx.doi.org/10.1111/jsm.12391] [PMID: 24251665]

[195] Mattagajasingh I, Kim CS, Naqvi A, *et al.* SIRT1 promotes endothelium-dependent vascular relaxation by activating endothelial nitric oxide synthase. Proc Natl Acad Sci USA 2007; 104(37): 14855-60. [http://dx.doi.org/10.1073/pnas.0704329104] [PMID: 17785417]

[196] Zemel MB, Kolterman O, Rinella M, *et al.* Randomized Controlled Trial of a Leucine-Metformi--Sildenafil Combination (NS-0200) on Weight and Metabolic Parameters. Obesity (Silver Spring) 2019; 27(1): 59-67. [http://dx.doi.org/10.1002/oby.22346] [PMID: 30569637]

[197] Hughes TE, Kim DD, Marjason J, Proietto J, Whitehead JP, Vath JE. Ascending dose-controlled trial of beloranib, a novel obesity treatment for safety, tolerability, and weight loss in obese women. Obesity (Silver Spring) 2013; 21(9): 1782-8. [http://dx.doi.org/10.1002/oby.20356] [PMID: 23512440]

[198] Elfers CT, Roth CL. Robust Reductions of Excess Weight and Hyperphagia by Beloranib in Rat Models of Genetic and Hypothalamic Obesity. Endocrinology 2017; 158(1): 41-55. [http://dx.doi.org/10.1210/en.2016-1665] [PMID: 27849360]

[199] Kopelman P, de Groot HG, Rissanen A, *et al.* Weight loss, HbA1c reduction, and tolerability of cetilistat in a randomized, placebo-controlled phase 2 trial in obese diabetics: comparison with orlistat (Xenical). Obesity (Silver Spring) 2010; 18(1): 108-15. [http://dx.doi.org/10.1038/oby.2009.155] [PMID: 19461584]

[200] Takagi K, Legrand R, Asakawa A, Amitani H, François M, Tennoune N, *et al.* Anti-ghrelin immunoglobulins modulate ghrelin stability and its orexigenic effect in obese mice and humans. Nature Communications 2013; 4(1): 1-11. [http://dx.doi.org/10.1038/ncomms3685]

[201] Haffer KN. Effects of novel vaccines on weight loss in diet-induced-obese (DIO) mice. Journal of Animal Science and Biotechnology 2012; 3(1): 1-7.

SUBJECT INDEX

A

B

C

Shazia Anjum (Ed.)

D

E

F

G

H

I

K

L

M

N

O

P

R

S

T

U

V

W

Z

Made in the USA
Las Vegas, NV
30 April 2026

7b74c445-3207-4742-bb4c-cc7d2d6c5547R04